Research Reports ESPRIT

Subseries
Project 322 · CAD Interfaces (CAD∗I)
Volume 5

Subseries Editors:
I. Bey, Kernforschungszentrum Karlsruhe GmbH
J. Leuridan, Leuven Measurement and Systems

Edited in cooperation with
the Commission of the European Communities

W0042039

M. Raflik B. Pätzold (Eds.)

CAD∗I Database

An Approach to an Engineering Database Version 4.0

With Contributions by

G. Feldmeier (BMW)
R. Grandl (BMW)
H. Libouban (CISIGRAPH)
B. Pätzold (UKA)
M. Raflik (CISIGRAPH)
W. Weick (KfK)

Springer-Verlag
Berlin Heidelberg GmbH

Volume Editors

Michel Raflik
CISIGRAPH, Data Base Group
Les Bureaux de Parc la Griffon
590, Route de la Seds, F-13127 Vitrolles, France

Bernd Pätzold
Universität Karlsruhe
Institut für Rechneranwendung in Planung und Konstruktion
Kaiserstr. 12, W-7500 Karlsruhe, FRG

ESPRIT Project 322: CAD Interfaces (CAD∗I) belongs to the Research and Development area "Computer-Aided Design and Engineering (CAD / CAE)" within the Subprogramme 5 "Computer-Integrated Manufacturing (CIM)" of the ESPRIT Programme (European Strategic Programme for Research and Development in Information Technology) supported by the European Communities.

ESPRIT Project 322 has been established to define the most important interfaces in CAD / CAM systems for data exchange, data base, finite element analysis, experimental analysis, and advanced modelling. The definitions of these interfaces are being elaborated in harmony with international standardization efforts in this field.

Partners in the project are:
Bayerische Motorenwerke AG / FRG · CISIGRAPH / France · Cranfield Institute of Technology / UK · Danmarks Tekniske Højskole / Denmark · Estudios y Realizaciones en Diseño Informatizado SA (ERDISA) / Spain · Gesellschaft für Strukturanalyse (GfS) mbH / FRG · Katholieke Universiteit Leuven / Belgium · Kernforschungszentrum Karlsruhe GmbH / FRG · Leuven Measurement and Systems / Belgium · NEH Consulting Engineers ApS / Denmark · Rutherford Appleton Laboratory / UK · Universität Karlsruhe / FRG.

ISBN 978-3-540-53383-2   ISBN 978-3-642-84335-8 (eBook)
DOI 10.1007/978-3-642-84335-8

This work is subject to copyright. All rights are reserved, whether the whole or part of the material is concerned, specifically the rights of translation, reprinting, re-use of illustrations, recitation, broadcasting, reproduction on microfilms or in other ways, and storage in data banks. Duplication of this publication or parts thereof is only permitted under the provisions of the German Copyright Law of September 9, 1965, in its version of June 24, 1985, and a copyright fee must always be paid. Violations fall under the prosecution act of the German Copyright Law.

Publication No. EUR 12937 EN of the
Commission of the European Communities,
Scientific and Technical Communication Unit,
Directorate-General Telecommunications, Information Industries and Innovation,
Luxembourg
Neither the Commission of the European Communities nor any person acting on behalf of the Commission is responsible for the use which might be made of the following information.

© 1990 Springer-Verlag Berlin Heidelberg
Originally published by ECSC - EEC - EAEC, Brussels - Luxemborg, in 1990

2145/3140 – 543210 – Printed on acid-free paper

CAD*I Project Overview

During the past 25 years computers have been introduced in industry to perform technical tasks such as drafting, design, process planning, data acquisition, process control and quality assurance. Computer-based solutions, however, are still in most cases single isolated devices within a manufacturing plant.

Computer technology is evolving rapidly, and the life cycle of today's products and production methods is shortening, with continuously increasing requirements from customers, and a trend to market interrelations between companies at a national and international level. This forces a growing need for efficient storage, retrieval and exchange of information. Integration of information is urgent within companies to interconnect departments which used to work more or less on their own. On the other hand direct communication with outside customers, suppliers and partner institutions will often determine the position of an enterprise among its competitors. In this sense, Computer Integrated Manufacturing (CIM) is the key of today for the competitiveness of tomorrow. But the realization of a future-oriented CIM concept is not possible without powerful, widely accepted and standardized interfaces. They are the vital issue on the way to CIM. They will contribute to harmonizing data structures and information flows and will play a major role in open CIM systems. Standardized interfaces allow:

- Access to data produced and archived on computing equipment which is no longer in active use;
- Communication between hardware and software from different vendors;
- Paperless exchange of information.

ESPRIT Project 322 "CAD Interfaces" (CAD*I) started in 1984 is a five-year research and development programme on CAD interfaces with the aim of defining some missing interface specifications in the environment of computer aided design (CAD) systems for mechanical engineering. Parts design and CAD are the starting point in the design and manufacturing process, and can also be considered as a starting point for information generation and data exchange.

Based on the results and using the experiences of former national standardization initiatives like IGES, VDAFS or SET, the CAD*I project team aimed from early in the project to contribute to the first international standard for product data exchange, because only an internationally accepted standard interface will fulfill the requirements of European industry.

The standardization work in CAD data exchange at international level is performed through ISO/TC184/SC4 under the name STEP: Standard for the Exchange of Product Model Data. CAD*I has had a large influence on the STEP definitions especially for the exchange of geometry and shape information (curves, surfaces and solid models), the interface to Finite Element Analysis programmes and drafting information.

This report is one of a series of similar books which summarize the wealth of results achieved during the five years of ESPRIT Project CAD*I.

CAD Interfaces

The main results are:

- Vendor independent interface consisting of a neutral file specification and corresponding pre- and post-processors for many commercial CAD systems have been defined, developed and tested. The CAD*I specifications for geometry and shape representation (curves, surfaces and solids) are clearly visible in the first international draft proposal standard. The processors are in practical use in several European and national projects. European system vendors have begun to integrate these results into their products.

- A general standard specification of a neutral file for exchanging finite element data has been developed and implemented. Tests have been performed with the interface processors for several FEM packages available on the market. In addition CAD models were transferred to finite element systems using the CAD*I neutral file. The results of this work have already appeared on the European market.

- New and improved data acquisition and analytical procedures for dynamic structural analysis have been specified and tested on complex real structures. Also, powerful tools for the intelligent integration (link) of experimental and analytical results in structural design have been developed, tested and merged into software products now available on the market. These results are visible in recent commercial products.

- Some new methods have been developed to enhance the communication interface in CAD/CAE systems. Future users of this kind of system will be able to enter information to the systems by handsketching input or by technical terms from using design language instead of via formal geometrical descriptions. First implementations have been successful; they are based on levels of internal interfaces using a product model.

- A neutral database interface based on the CAD*I neutral file format has been developed to handle archiving and retrieval of product information in a database. A set of standard access routines has been written and tested with existing CAD systems and a widely used commercial relational database management system. The introduction of these results into marketable products is on the way.

- An information model for the description of technical drawings has been developed: the CAD*I drafting model. This information model represents the highest level of sophistication within the level concept of the drafting model of the STEP specification.

A total of about 150 person-years of research and development effort has been spent on the project. The CAD*I project involved 12 partners in 6 countries of the European Community.

As project manager since 1985 I would like to express my appreciation to the co-manager J. Leuridan and the fifty or more people working in and on the project for their engagement to reach the originally stated goals. In addition I would like to pay special tribute to Mrs. P. MacConaill and R. Zimmermann form the Commission of the European Communities and to the reviewers G. Enderle (+), E.A. Warman and H. Nowacki for their cooperative support. Special acknowledgement is due also to Mrs. U. Frey for running the administrative part of the project and for her contributions to forming the spirit of the CAD*I team.

I. Bey, CAD*I Project Manager

Table of Contents

1 Introduction

Companies produce various types of data (administrative, technological and other) which are more and more often stored in a database. This enables any authorized user:

☞ to know what data produced in the company is available to him or her

☞ to benefit from the results of the other users

☞ to be informed of any modification made by any user

☞ to communicate better with users from other departments

and so on.

However, these functionalities cannot be developed in the CAD area because of the difficulty of managing easily the isolated islands that are the CAD data stored in independent and insufficiently flexible "files".

It is a nearly impossible job to create and maintain relations between pieces of information (entities) of one model stored in one "file" and pieces of information (entities) of another model stored in another "file".

The aim of CAD*I Database is to demonstrate that a Database System is the appropriate tool to create, maintain and modify these complex relations. In that way, CAD*I Database specified and implemented a standardized communication interface with a Database System. This communication interface consists of a set of standard subroutines for Fortran applications to write, to read, to modify, to delete, etc. CAD data in a Database.

Since users use DBMS to store the definition of complex products, and since administrative information (in particular access rights and management of versions) have a major importance for complex products CAD*I Database considered it to be of great importance to include in the CAD*I data schema the most commonly administrative data types used in industry.

Therefore in addition to the geometry data defined by the project WG1, WG2 and WG3, one of the aims of CAD*I Database is to specify and implement the standard subroutines which will operate on the administrative and access control data included in the CAD*I data schema.

1.1 Basic Principles

1.1.1 Problems Met by CAD Users When Developing Programs Using CAD Data

More and more CAD users (such as the automobile and aerospace industries) develop application programs specific to their industry, company, or department, based on CAD data.

These application programs represent for these users both a considerable effort of development and an increasing part of their know-how stored in the computer. Most of the time, these application programs are dependent on a specific CAD system. Indeed, they access (write, read...) CAD data via a library of subroutines provided by their CAD vendor. Since the specifications of these subroutines are not standardized, the application programs are strongly dependent on this library. Furthermore, it is very difficult for these users to obtain the source code of this library from their CAD vendor (it represents their know-how), and when they can buy it, most of the time it is computer-dependent (for CPU efficiency).

This situation binds these CAD users strongly both to a specific CAD system and to a specific computer.

1.1.2   Problems Met by CAD Developers When Developing CAD Interfaces

CAD data schemas proposed by standardization organizations are becoming more and more complex.

The range of applications is permanently increasing (to cover solids, surfaces, drawings, finite elements, numerical control, piping, schematic applications, and so on) and the capabilities in each application are getting more and more sophisticated (with new concepts, parameterization and suchlike).

Therefore, developing an interface software that:

☞   performs well (in terms of CPU time)

☞   covers the whole domain of information defined by the standard

☞   is flexible enough to integrate new specifications of the standard

☞   is able to make efficient consistency checks on neutral data

requires an increasing investment from CAD suppliers.

It is clear that if some parts of software needed by all CAD developers to develop their interfaces could be specified and developed only once, it would represent a great benefit both for CAD vendors and CAD users. In particular, a library of subroutines to access (write, read...) CAD data stored according to the neutral format would be very helpful for :

☞   easily accessing data,

☞   checking data consistency.

1.1.3   What is Standardized in the CAD Field Today?

Today, standardization organizations propose:

☞   Conceptual level specifications of a CAD data schema

☞   Physical level specification of this CAD data schema in a sequential ASCII file

On the contrary, at present , there is no standardization of the specifications of a library of subroutines to access the CAD data stored according the neutral format.

Remark:

The "Aerospatiale" company has proposed a package of subroutines named "XSET" to access (read and write operations) data in a SET "NEUTRAL FILE" (Fig. 1.1). However, this library is not a standard at present.

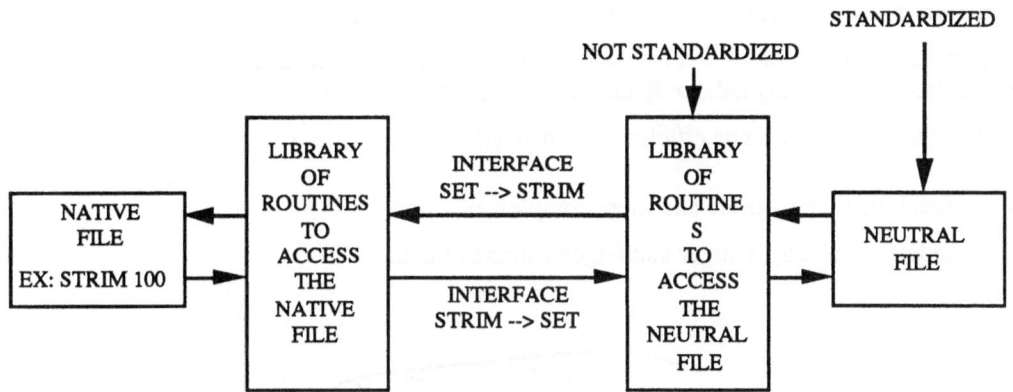

Fig. 1.1   Software structure of a CAD interface

(Example: STRIM 100 and SET)

1.2 The CAD*I Database Proposal

a) The CAD*I database group proposes to take one more step in the standardization by specifying and standardizing a library of subroutines to access CAD data stored according to the CAD*I Neutral Format.

The specifications of standard subroutines are independent of the physical format of data stored. Therefore the subroutines can be implemented equally on:

-   a sequential ASCII file

-   a hierarchical database

-   a relational database

-   and in other ways.

b) The CAD*I database group decided to implement a subset of this package on a relational database system offering the standard SQL language.

Such a choice is closely related to :

-   the capability of the standard SQL language to describe flexible and powerful functionalities

-   the capability of the relational DBMS to execute these functionalities with a high level of performance

The CAD*I database group wants indeed to offer a library of flexible subroutines able to manage complex objects (complex relations + large volume of data) efficiently (measured in terms of CPU time).

Thanks to its parametrization functionalities, the CAD*I data schema has a high capability to define complex relations between objects.

For example shape constraints can be defined between different objects (common radius, common boundary defined by reference to the same curve...).

Figure 1.2 illustrates the actual limitation of CAD systems to manage very large volume of data efficiently (in this model only the skin of the car is defined).

In order to achieve the same efficiency in managing:

-   models containing more and more complex relations,
-   models defined by a larger and larger volume of data,

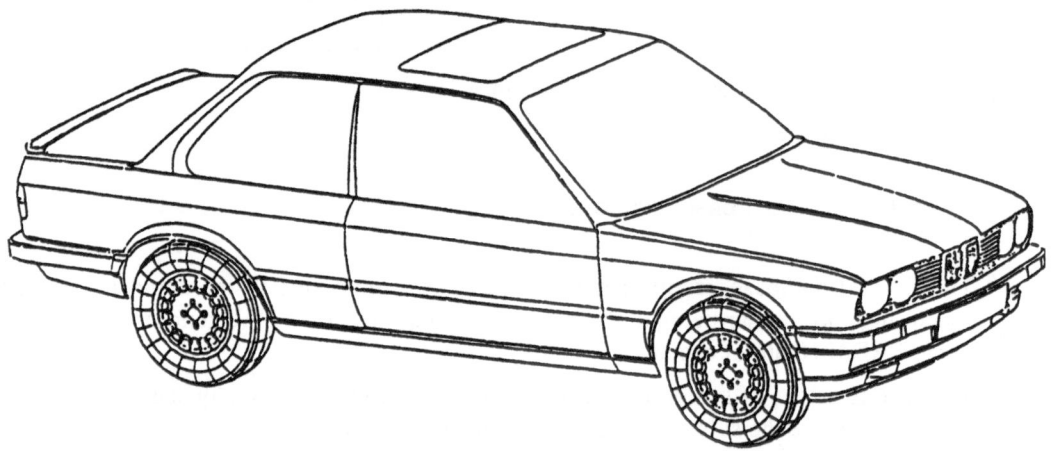

Fig. 1.2   Complex part

CAD developers need to develop a software package to access CAD data. If they continue in this direction they will develop a CAD data access software package that has a complexity more and more comparable with the complexity of a DBMS.

To this end, the implementation of the standard library must be done with a modern and powerful tool: a database management system offering the standard SQL language. The tests on the prototype developed will be done only with the database system ORACLE.

This data base system was chosen for the project for the following reasons:

-   it is a relational database system
-   it has an SQL interface
-   it is portable (between IBM, Digital, and Bull, for example)
-   it has many references in industry.

c) To take one more step in order to represent an ever increasing volume of data, CAD*I database group considers that it is necessary to complete the CAD*I data schema with some administrative information.

Indeed in a database context, geometric information is not sufficient to describe complex objects, and the management of more complex products (many parts defined by different CAD users) requires control of the access to this data. In addition, selection criteria are object-oriented criteria and not geometrical ones.

Thus, one of the aims of CAD*I database group is to identify and to include in the CAD*I data schema some of the most common administrative data used in industry, for example :

- information on access rights

- records of design versions

- data specifying design variants

- description of project organization (listing, say, groups or users)

- a record of who is actually modifying an object

- and similar data

The main tasks of the CAD*I database group are :
- specification of the administrative data types to be included in the CAD*I data schema

- specification of the CAD*I standard subroutines to access the CAD*I data in a database

- implementation of these subroutines for relational database systems via the standard SQL language

- specifications of the functionalities and subroutines to access the administrative data

- implementation of these subroutines for relational database systems via the standard SQL language

- development of a demonstration program able to exchange CAD data between different CAD systems via a relational database system and based on the implementations of :

    - a neutral application Interface (AIS) <--> CAD*I-ORACLE Interface software

    - the STRIM100 CAD system <--> CAD*I-ORACLE Interface software

    - the CAD*I NEUTRAL FILE <--> CAD*I-ORACLE Interface software.

1.3 The CAD*I Database Group Achievements

The main realizations performed in the CAD*I database group are the following:
- Specification and implementation on the Relational DBMS ORACLE of the CAD*I database interface subroutines related to the management of :

    - the subset of the CAD*I geometric data structure related to B_REP models with planar faces,

    - the CAD*I Administrative data structure defined and implemented by the CAD*I database group (Administrative data, Access rights, Closed model,etc)

- Specification and implementation of the following interfaces to allow CAD exchanges between different CAD systems accessing CAD*I data stored in the CAD*I Neutral file or the CAD*I-ORACLE database :

  - AIS - Modeller --> CAD*I-ORACLE

  - STRIM100 CAD system --> CAD*I-ORACLE

  - CAD*I-NEUTRAL FILE --> CAD*I-ORACLE

- Specification and implementation of the CAD*I Database Administrative Data Manager the aim of which is to manage CAD*I data stored in the CAD*I-ORACLE database and CAD*I data exchange between different CAD systems.

1.4 Expected Benefits for the Users of the CAD*I Subroutines

1.4.1 Benefits from Standardizing Routines Specifications

a) CAD users can develop their specific application programs independently of any CAD system and any computer, based on:

- The standard data schema,

- The standard "Neutral Interface".

b) CAD developers will spend less energy in developing, maintaining, and completing their interface softwares or application programs using the Neutral Interface.

Indeed, the Neutral Interface subroutines are developed only once and are immediately available to application programs to access CAD*I data.

c) The reliability of CAD data exchange will be improved. At present a lot of time is lost finding out which of the sending or receiving system is responsible for transfer errors. The Neutral Interface can check (at least partially) the consistency of the neutral data to be stored.

For example the routine writing in the Neutral File of an IGES Linear Dimension Entity can check whether the intersection between the Dinension Line 1 and the Witness Line 1 of the Linear Dimension Entity (see Fig. 1.3) corresponds to the arrowhead.

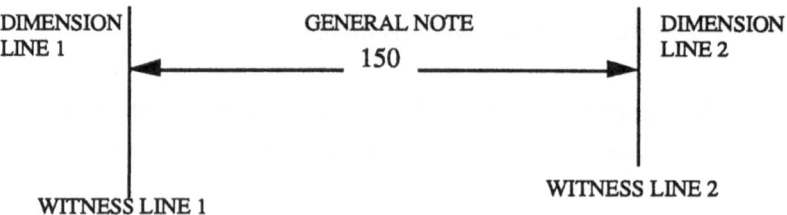

Fig. 1.3   Linear dimension entity (IGES)

Thus, if all applications use the same subroutines (specified and implemented by standardization organizations) to write the neutral entities, the number of incoherent Neutral Files would decrease considerably.

d) The precision of the data schema specifications may increase progressively.

Some terms like "continuity" are used by standards like IGES, SET, VDA but are vaguely defined. "Continuity" requires the definition of tolerance and criteria (applied using an algorithm) to decide whether continuity is respected or not.

For example the safest way to create a file that contains a surface whose patches respect continuity is to call a standardized subroutine (specified and implemented by standardization organizations) that writes a surface and returns an error code if continuity is not sufficient according to the standard.

1.4.2 Benefits from Implementing on a Relational Database System

a) With a DBMS we ensure the applications and operations available today are well managed.

At present, in addition to the usual POSTPROCESSOR application using a read operation, and PREPROCESSOR application using a write operation, the user application programs using read, write, update and delete operations are increasingly being developed within companies (see Fig. 1.4).

However, such operations are difficult to implement with classic languages and sequential ASCII file techniques (the most often used at present).

A database management system which allows the handling of a large volume of data, and the associated standard SQL language which has great flexibility for read, write, update and delete operations are the appropriate tools for all the programming applications.

For example, independently stored objects can be related, later on, through common parameters that define geometrical constraints. A DBMS is very powerful for update operations. These parameters can be numerical parameters or geometrical parameters (such as line or curve). If, say, an object is stored with a reference to a parameter R1 (radius) and a second object is stored with a reference to a parameter R2 (radius) and then the need appears for a common value for the radius of both objects, then the reference to parameter R2 can be replaced by a reference to parameter R1.

b) The implementation of the standard subroutines is valid for

- any computer with FORTRAN compiler
- any relational DBMS communicating via the SQL language.

Indeed the Neutral Interface subroutines are composed of:

- standard FORTRAN statements
- standard SQL statements.

	PREPROCESSOR	POSTPROCESSOR	USER APPLICATIONS
READ ENTITIES		X	X
WRITE ENTITIES	X		X
MODIFY ENTITIES			X
DELETE ENTITIES			X

Fig. 1.4   The operation types executed by the different applications

c)   A relational database system ensures security and transparency for the users of CAD data since:

CAD data stored in a relational database are accessible to CAD users via the standard SQL language. CAD users can therefore be informed of the evolution of the implemented data schema.

d)   A database system improves communication between departments within companies :

Data produced is stored in a centralized database in a standard format. Therefore:

-   the CAD data produced in the company are immediately available

-   the modifications made on CAD data by any authorized CAD user are available for all authorized CAD users as soon as they are achieved.

e)   Storage of CAD data in a database system is a prerequisite for the development of sophisticated functionalities:

-   Storing different versions of the same object without duplicating information (common entities belonging to the same object can be referenced many times).

-   Creation of sophisticated assemblies and selection of a subset (to see the placing of a dummy in a car, it is sufficient to select only the geometrical shell of the car and then check that they fit together).

-   To retrieve data according to any geometrical or relational selection criteria (to study how to adjust a door to a side, a CAD user selects only the relevant entities).

-   To allow the definition of complex relations between objects including parametrization possibilities.

1.4.3 Benefits from Adding Administrative Data in the CAD*I Data Schema

The subset of administrative data improves security, communication and productivity within companies :

The administrative data protects each geometry against illicit access. Indeed every user is able to define access rights to his own geometry. Administrative information is available on any object stored in the database. (for example, on the status of the object, which version it is, which variant, its owner, who is modifying the object , etc.)

A sophisticated management of versions can be described. It is possible to know exactly:

☞ which geometrical entities were modified between two versions of a same object

☞ who made the modification

☞ when the modification was made

☞ why the modification was made in case of modification of the geometry of an object whose unchanged geometric entities are not duplicated.

2   The Subset of the CAD*I Data Schema

In the CAD*I Database Group we handle geometry in the form of closed models as well as in the form of models divided into their single points, lines, directions, planar surfaces and topology. A closed model contains geometry in any format (in a second step also a part program or NC-data). It is impossible to look into the closed model, to recognize any structure in it. This view of data is used to archive and exchange data via a common database. To inform about the content, each closed model possesses a set of administrative data.

Also each assembly in the database must be described by a set of administrative data, which means administrative data can have a hierarchical structure (see Figs. 2.2 - 2.17): an assembly is recursively built up by other assemblies and B_Reps (see the CAD*I Specification of a "Neutral File for Solids"). The lowest level possessing a set of administrative data is the B_Rep or the closed model. If various B_Reps are collected into an assembly, each B_Rep possesses its own set of administrative data and in addition the whole assembly as a unit possesses a set of administrative data, which references the administrative data of the corresponding B_Rep. The term assembly is used only in connection with part or tool geometries. The same procedure is valid for assemblies consisting of other assemblies and B_Reps.

Summarizing, administrative data are used:

☞   to describe a model

☞   to control access to a model

☞   to build classes of models.

With the administrative data the technical designer can inform others, or be informed, about what is in the database, in which status, by whom created, wether it is valid or not, whether there already exists already a new version, and so on. There exist manifold possibilities of inquiry.

This specification is set up such that for each entity, property, or attribute we define in sequence

1.   the conceptual schema in IDEF1x

2.   the data structure formally in HDSL

3.   the semantic description in textual form (sometimes with illustrations). The attribute names used in the HDSL definition are used in the text and in the illustrations.

For each entity, class, property, relation, and attribute type in the formal schema definition in IDEF1x and HDSL we will now give an informal description of its meaning. This will be done by describing the individual attributes of the respective data structures.

2.1 Conceptual Schema in IDEF1x

In the following the conceptual schema of the partial CAD*I reference model which will be briefly described in IDEF1x. The concept of the IDEF1x modeling technique is based on the Entity-Relationship approach. A glossary of the IDEF1x symbols is given in Fig. 5.1.A detailed description is given in /ISO-N273/.

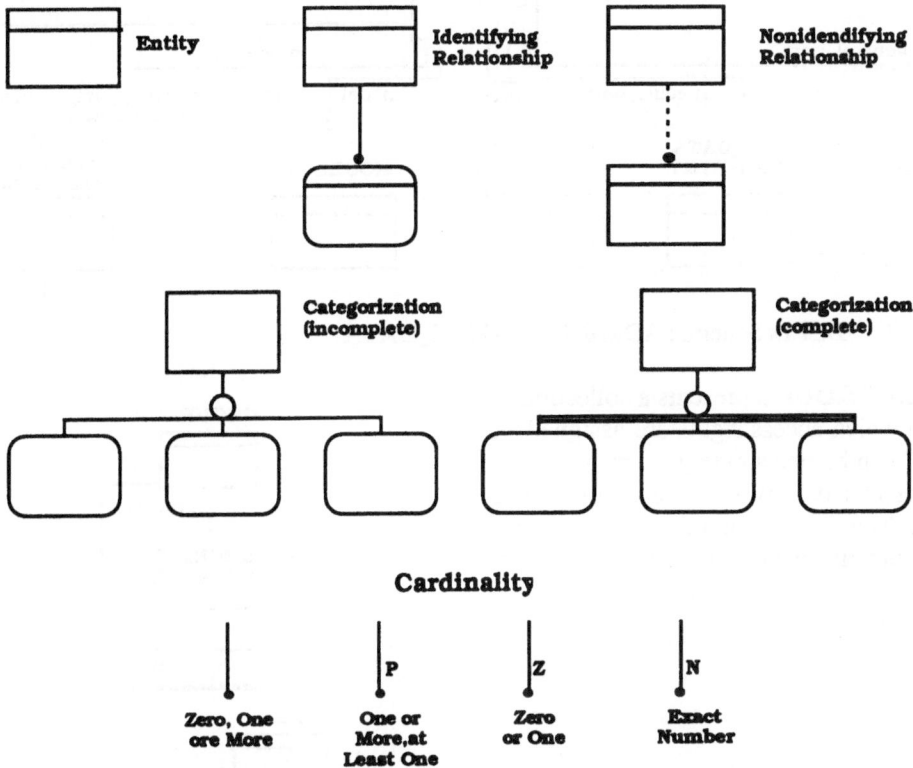

Fig. 2.1 Glossary of IDEF1x symbols

The entity **ADM_DATA** represents the classification of the administrative data.

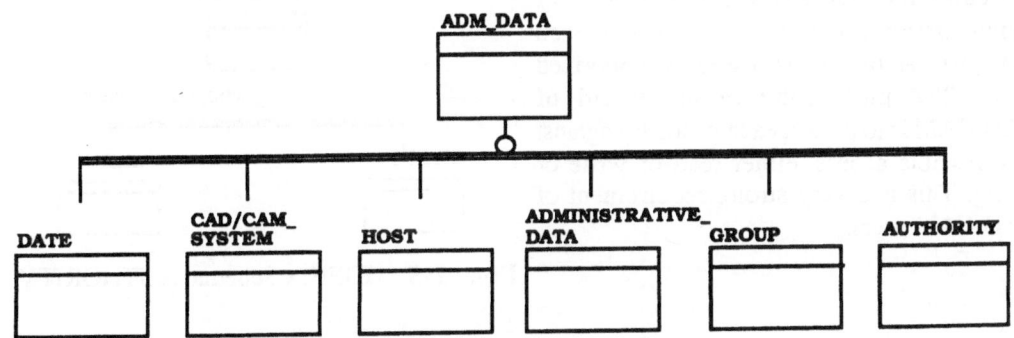

Fig. 2.2 IDEF1x Schema: ADM_DATA

The entity **ADMINISTRATIVE_DATA** represents a verbal, informal description of each model (and its environment) that has to be administered in the database. Each closed model, each assembly (a hierarchy of assemblies, R_Reps and closed models containing geometry) and each B_Rep must be described by a set of administrative data.

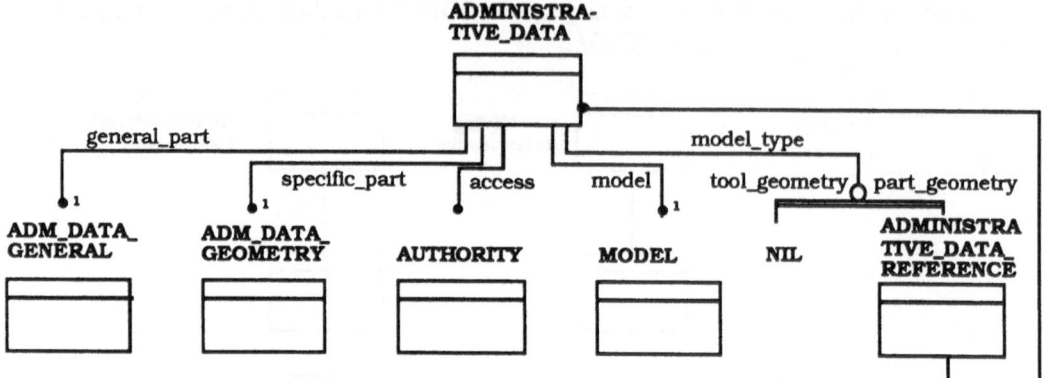

Fig. 2.3   IDEF1x Schema: ADMINISTRATIVE_DATA

The entity **GROUP** represents a collection of users to which access rights are assigned. A GROUP can be one single user or one single group or can be any mixture of users and groups. The definition of a group is recursive, but may not contain circles.

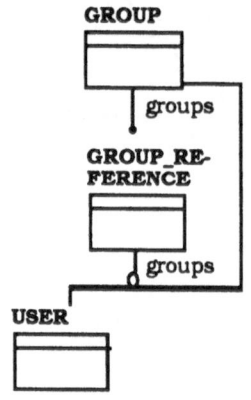

Fig. 2.4   IDEF1x Schema: GROUP

The entity **AUTHORITY** describes a set of administrative data according to which each geometry can only be accessed by authorized users. The philosophy in the world of CAD/CAM is to protect each geometry against inadmissible access, either read or write or delete. This is a very strong requirement of CAD/CAM users.

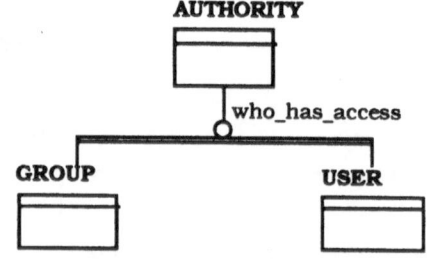

Fig. 2-5   IDEF1x Schema: AUTHORITY

The entity **ADM_DATA_GENERAL** represents the general part of the administrative data, that contains attributes of general validity, necessary for all types of models.

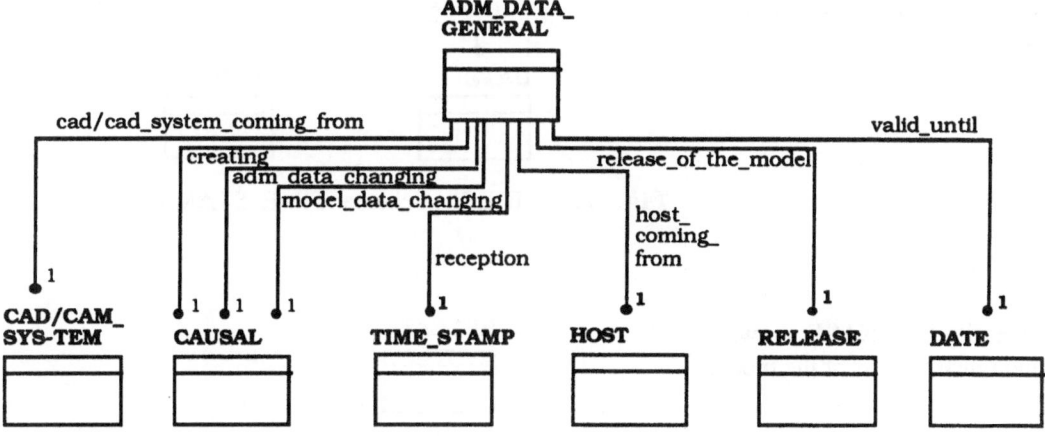

Fig. 2.6   IDEF1x Schema: ADMINISTRATIVE_DATA_GENERAL

The entity **GEOMETRIC_- MODEL** is a class representing entities according to the geometric modeling techniques For the CAD*I data base we will only use the SOLID_MODEL and the technique of CLOSED_MODEL.

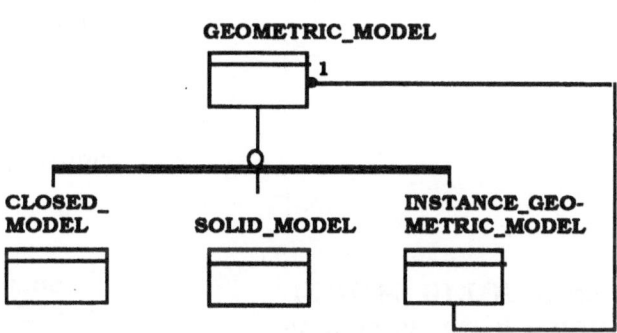

Fig. 2.7   IDEF1x Schema: GEOMETRIC_MODEL

The entity **CAUSAL** gives information about an event for example, an executed change or creation of data .

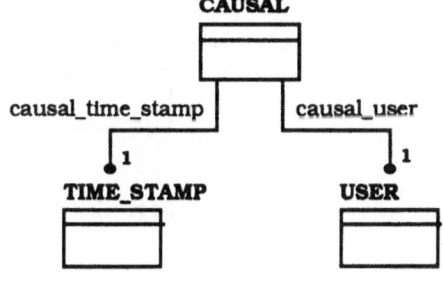

Fig. 2.8   IDEF1x Schema: CAUSAL

The pair of entities **date** and **time** together from the entity **TIME STAMP**, to mark special events, such as creation or change of data.

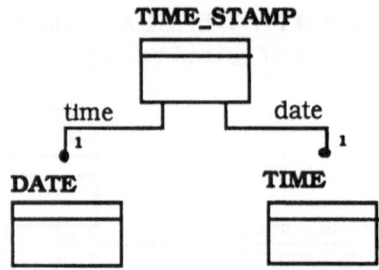

Fig. 2.9   IDEF1x Schema: TIME_STAMP

The entity **RELEASE** shows the valid release status of the model.

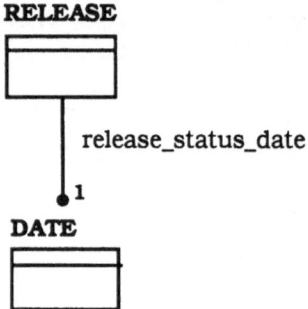

Fig. 2.10 IDEF1x Schema: RELEASE

The entity **SOLID_MODEL** describes a B_rep model or an instance of a solid model.

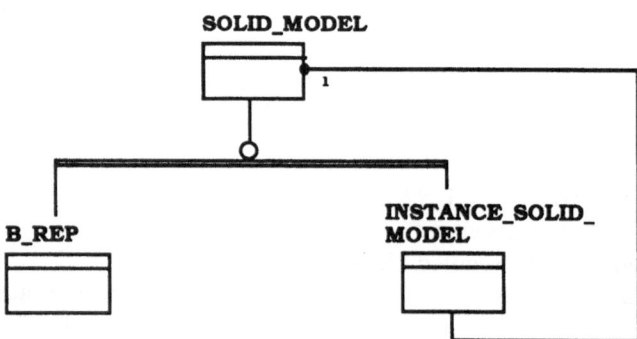

Fig. 2.11 IDEF1x Schema: SOLID_MODEL

A **B_REP** is an entity that has a scope. A B_REP may have a material property associated with it. The B_REP is a 'self-contained' entity in the sense that no entity in the B_REP may refer to an entity outside the scope of the B_REP. All referenced entities must be within the B_REP scope itself. The scope of a B_REP contains both topological and geometrical entities. Geometry is represented by lists of the entities POINT, DIRECTION, LINE and PLANAR_SURFACE which are referenced by the topological entities defined subsequently. Topology is represented by lists of the entities VERTEX, EDGE, EDGE_LOOP, FACE and SHELL in that order so that no entity is referenced before it is defined.

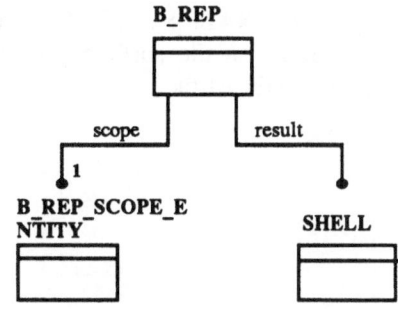

Fig. **2.12** IDEF1x Schema: B_REP

A **shell** is a topological entity defined within the scope of a B_REP. It is defined by a set of bounding faces. The faces must be connected and form a continuous surface which divides the three-dimensional space into two distinct regions.

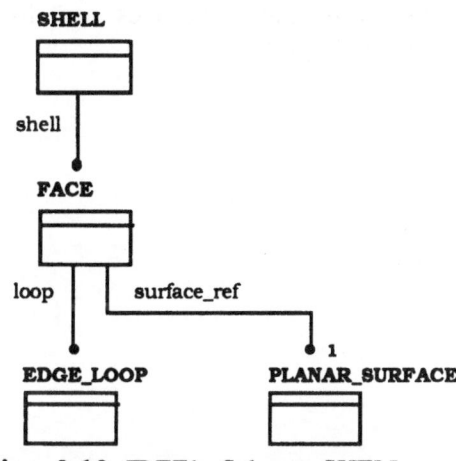

Fig. **2.13** IDEF1x Schema: SHELL

The entity **EDGE_LOOP** describes the boundary of a face by a set of edges or one vertex.

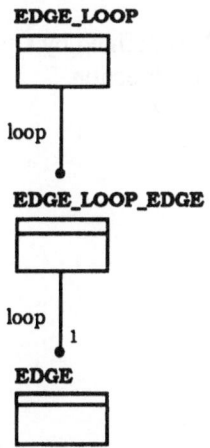

Fig. **2.14** IDEF1x Schema: EDGE_LOOP

The entity **PLANAR_SURFACE** is defined by a point on the surface and the normal direction to the surface.

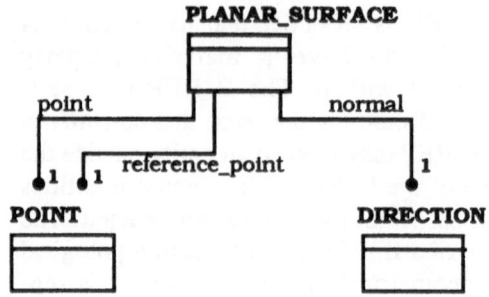

Fig. 2.15 IDEF1x Schema: PLANAR_-
          SURFACE

An **EDGE** is a topological entity defined within the scope of a B_REP. It is defined in terms of its bounding vertices and its underlying LINE geometry.

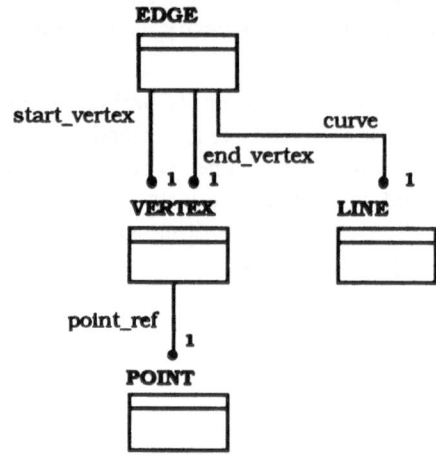

Fig. 2.16 IDEF1x Schema: EDGE

The entity **LINE** is defined by a point on the line and the normal direction.

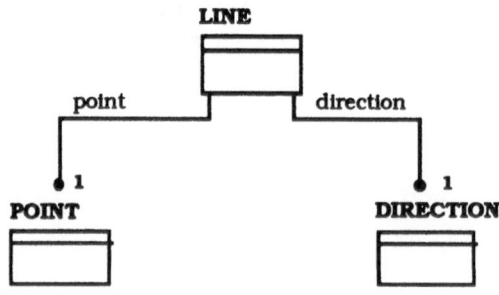

Fig. 2.17 IDEF1x Schema: LINE

2.2 Attribute Types and Reference Mechanism for General Use

The CAD*I reference model includes the following attribute types and referencing mechanisms.

2.2.1 Any

```
ATTRIBUTE ANY =            GENERIC (type: CLASS(        LINE,
                                                        POINT,
                                                        DIRECTION,
                                                        AUTHORIZED_USER,
                                                        USER_ID))
                           CLASS( type, REF_ANY(type) );
```

This is a class of attribute types which indicates that data is given either in elementary form or by reference to an entity of the corresponding type.

All entities which can appear in this form as attributes have two syntactical representations: one with a name as the first argument indicating that an entity of that type is to be created, and one without the name argument indicating that the data is to be treated as a structured attribute.

2.2.2 CAD/CAM-System

```
ENTITY CAD/CAM_SYSTEM = STRUCTURE
                                cad/cam_system_name :     STRING;
                                cad 9 :-9 : em_release :   STRING;
                        END;
```

The entity CAD/CAM_SYSTEM contains information about the CAD/CAM systems the model comes from. This information consists of the name and the release of the CAD/CAM-System, e.g. CATIA V2R2, STRIM V2R3, GILDAS V1R1, CD/2000 1.6.

2.2.3 Causality

```
ENTITY CAUSAL =            STRUCTURE
                                causal_time_stamp :       TIME_STAMP;
                                causal_user :             REFERENCE (USER);
                           END;
```

The entity CAUSAL gives information about an event, such as an executed change or creation of data .

Attribute name	Meaning and values
causal_time_stamp	Date and time the event (change or creation) took place
causal_user	The user identification of the user who caused the event

2.2.4 Date

ENTITY DATE =   STRUCTURE
                                  date :   string;
                          END;

where data has the format YYMMDD (Y= year, M = month, D = day).
For example December 6, 1987 must look like "871206".

2.2.5 Dim

ATTRIBUTE DIM = ENUM(D2, D2.5, D3) ;

This information does not appear explicitly on the neutral file. It is recognised from the fact that
either three or two or two-and-half coordinates are given. This attribute indicates two-, two-and-
half- and three-dimensional features.

2.2.6 Host

ENTITY HOST =   STRUCTURE
                                  host_type :   STRING;
                                  host_operating_system :   STRING;
                                  host_operating_system_release :   STRING;
                          END;

Attribute name	Meaning and values
host_type:	The name of the computer (e.g. VAX, IBM, CDC)
host_operating_system:	The operating system of the computer (e.g. VM/CMS, MVS, VMS)
host_operating_system_release:	The release of the operating system of the computer

2.2.7 Model Release

ENTITY RELEASE =   STRUCTURE
                                  release_status :   STRING;
                                  release_status_date :   DATE;
                          END;

The entity RELEASE shows the valid release status of the model.

Attribute name	Meaning and values
release_status	(partly) approved, (partly) checked up, released, experiment release, planning release, pre-production release, in production, etc.
release_status_date	The date of the last change of the release status

2.2.8 Ref_Any

ATTRIBUTE REF_ANY =   GENERIC (type: CLASS(PREDEFINED_ENTITY,
                                                        GEOMETRIC,
                                                        DIRECTION(DIM)))
                            CLASS(REFERENCE(type));

This is a class of reference types for referencing geometric and predefined entities. Such references may be internal (e.g. to entities transferred on the same file). References to formal parameters are allowed only within macros.

2.2.9 Time

ENTITY TIME =   STRUCTURE
                                    time :   string;
                            END;

where time has the format HHMMSS (H = hours, M = minutes, S = seconds).
For example "230000" means eleven o'clock at night.

2.2.10 Time Stamp

ENTITY TIME_STAMP =   STRUCTURE
                                    date :   DATE;
                                    time :   TIME;
                            END;

The pair consisting of date and time represents a time stamp, to mark special events such as creation or change of data.

2.2.11 User Identification

ENTITY USER =   STRUCTURE
                                    user_id :   STRING;
                                    user_name :   STRING;
                                    user_department :   STRING;
                            END;

A user in the CAD*I database is defined by his user identification, name and department. The user_id must be unique in the system. If necessary it will be generated.

2.3 Assembly Model

The assembly A (see Fig. 2.18) in this example consists of n B_Reps. Each B_Rep possesses a set of administrative data (level 1) which refers to the geometry of the B_Rep (level 0). The set of administrative data at the assembly level (level 2) refers to each set of adminstrative data of its B_Reps, but there is no direct reference to a geometry.

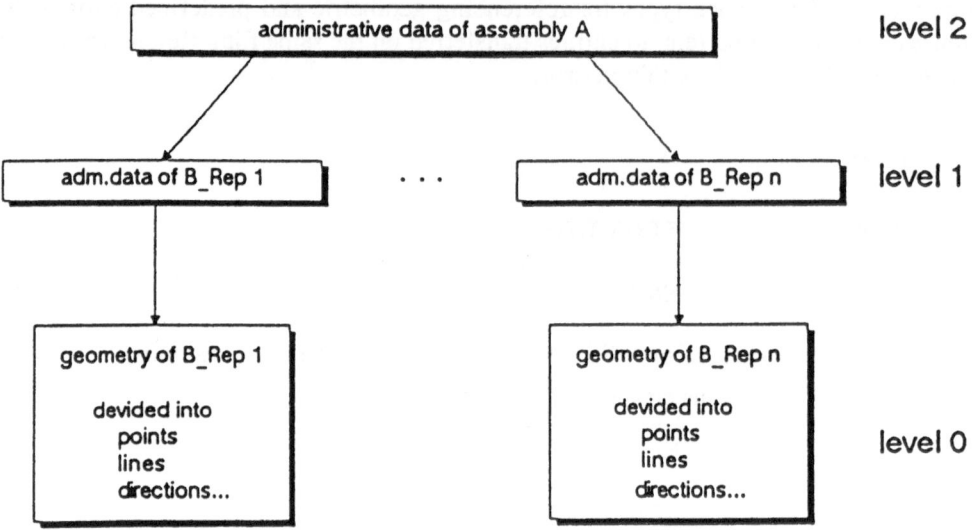

Fig. 2.18 Hierarchy of administrative data regarding assemblies

An assembly can also be used to collect closed models into a unit; in this case we have the following representation (Fig. 2.19):

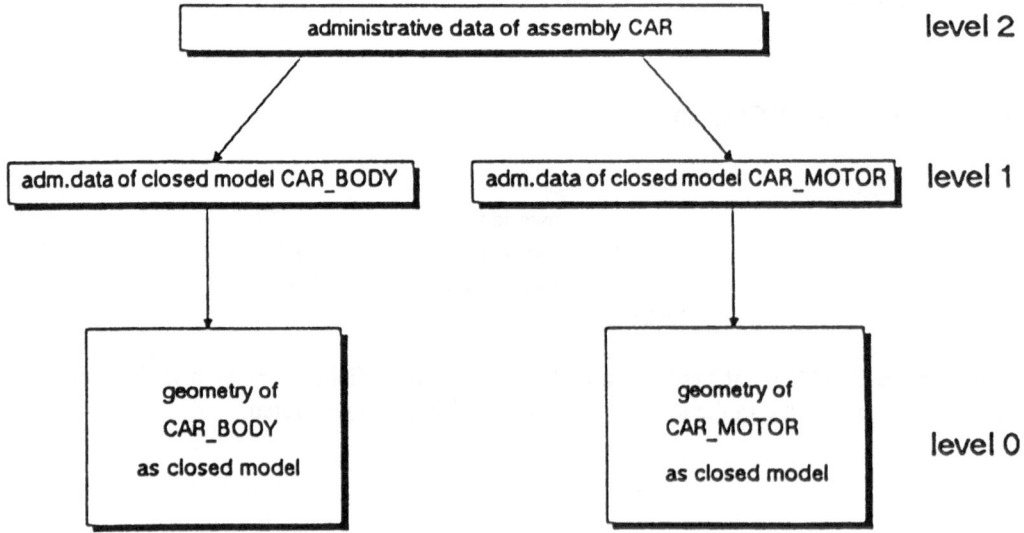

Fig. 2.19 Example of a hierarchy assembly

Regrading complex structures of assemblies (if an assembly itself consists of assemblies) data can exist in more than two levels, as shown in Fig. 2.20.

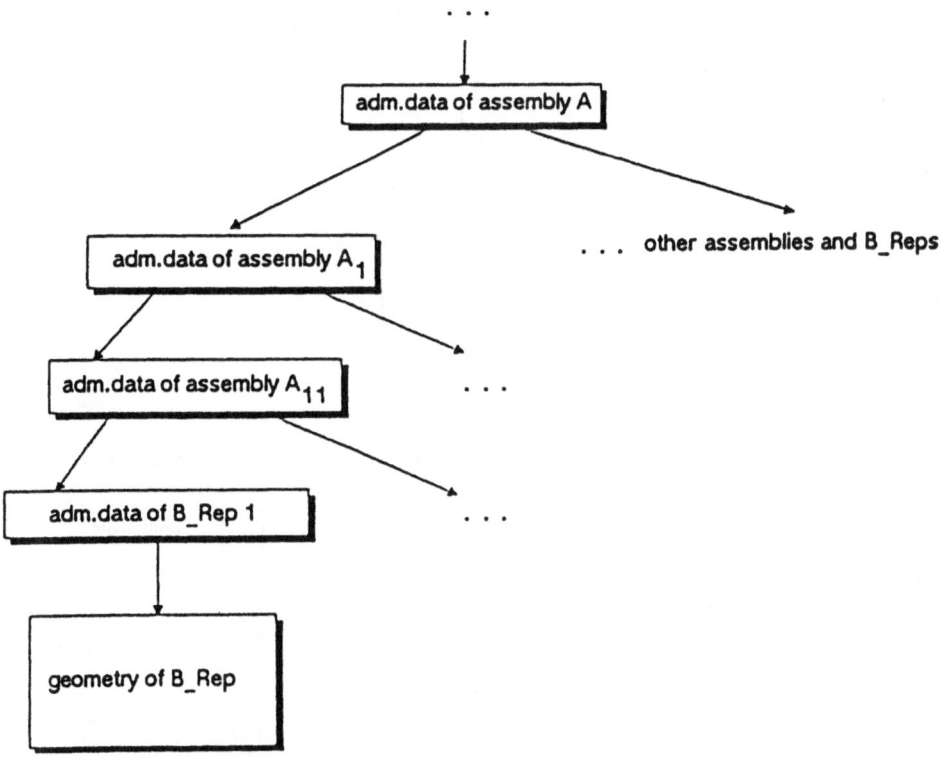

Fig. 2.20 Example of a complex assembly

If an assembly is an element (a member) of another assembly, its set of administrative data is referred to by the set of administrative data of the superior assembly.

Figue 2.21 defines the schema of the set of administrative data.

2.3.1 Type of Model

ENTITY MODEL_TYPE =   ENUM (PART_GEOMETRY, TOOL_GEOMETRY);

In the CAD*I database it is possible to store the geometry either of parts or of tools.

Depending on the type of the stored model, different statements or pieces of information are necessary to describe it.

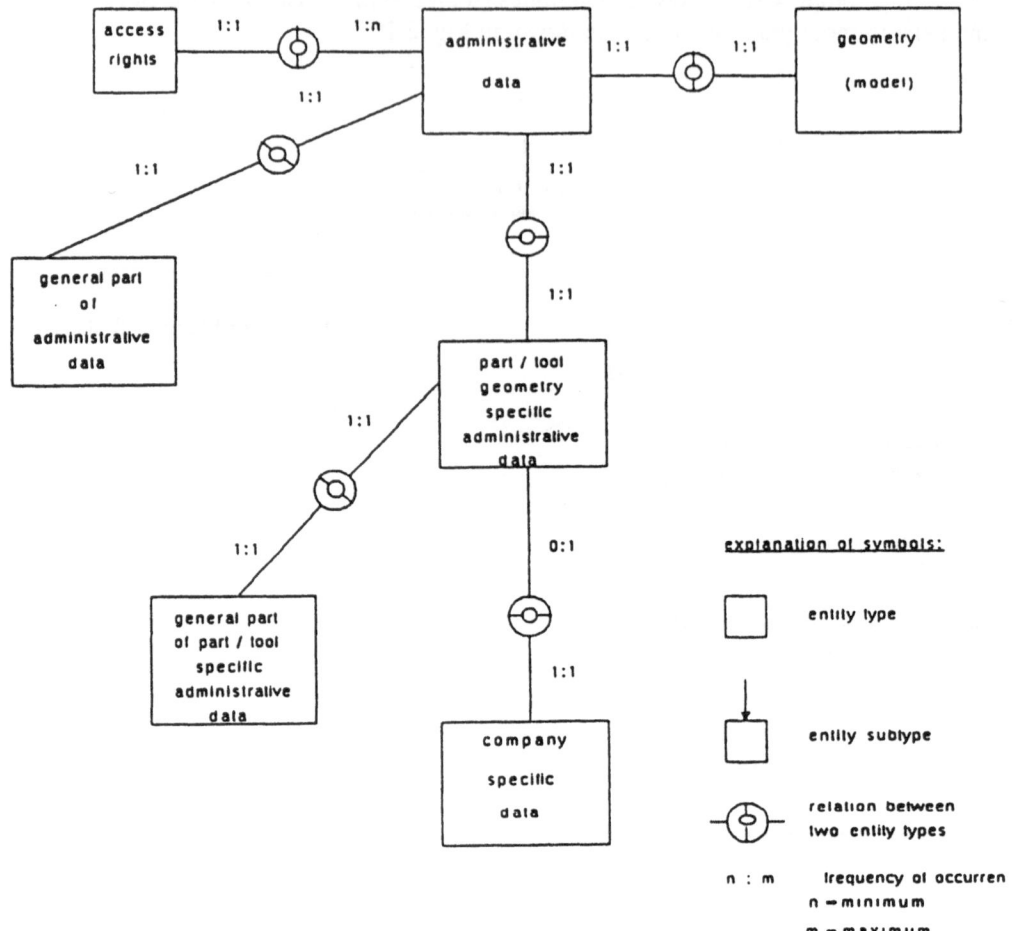

Fig. 2.21   Schema for the set of administrative data

2.3.2 Kind of Model

ENTITY MODEL =   CLASS (ASSEMBLY, GEOMETRIC_MODEL);

In the CAD*I database it is possible to store data either as a closed model (a unit without visible structure) or as a B_REP (structured geometry). A closed model can only ever be selected, inserted or deleted as a whole. It is impossible to look into it, and information about the contents can be obtained only via the corresponding administrative data. A closed model is used especially for archiving and data exchange and therefore can be of each type defined by the entity MODEL_TYPE. In a B_REP it is possible to select single points, lines and so on.

An assembly is a collection of other assemblies, B_Reps or closed models, which contain geometry. Each closed model, each assembly and each B_Rep possesses its own set of administrative data.

2.3.3   Class of Administrative Data

```
ENTITY ADM_DATA =          CLASS (    DATE,
                                      CAD/CAM_SYSTEM,
                                      HOST,
                                      ADMINISTRATIVE_DATA, GROUP,
                                      AUTHORITY)
```

The entity ADM_DATA represents the classification of the administrative data.

2.3.4   Administrative Data

```
ENTITY ADMINISTRATIVE_DATA =
                    GENERIC (model_type: MODEL_TYPE)
                    STRUCTURE
                              general_part :      ADM_DATA_GENERAL;
                              specific_part :     ADM_DATA_GEOMETRY;
                              access :            LIST OF REFERENCE(AUTHORITY);
                              model :             REFERENCE (MODEL);
                              CASE model_type OF
                                    PART_GEOMETRY : derived_model :
                                          LIST OF REFERENCE
                                                (ADMINISTRATIVE_DATA
                                                (TOOL_GEOMETRY));
                                    TOOL_GEOMETRY : derived_model : NIL);
                              END;
                    END;
```

The entity ADMINISTRATIVE_DATA represents a verbal, informal description of each model (and its environment) that has to be administrated in the database. Each closed model, each assembly (i.e. hierarchy of assemblies, R_Reps and closed models containing geometry) and each B_Rep must be described by a set of administrative data.

A set of administrative data can be divided into the following parts:

Attribute name	Meaning and values
general_part :	The general part of the administrative data contains attributes of general validity. They are necessary for all types of models.
specific_part :	The second part of the administrative data contains attributes that are specific to the type of the model. For example the set of administrative data belonging to a part geometry consists of the general part and the specific part that describes part geometries.
access :	Access to a model is only possible via its set of administrative data. The set of administrative data references a set of access rights saying who is allowed to read, write or delete this model (see Sect. 2.5).

| model : | This refers to the model described by the set of administrative data. |
| derived_model : | References are listed here to sets of administrative data for models directly derived from this model. |

2.3.5 General Part of Administrative Data

ENTITY ADM_DATA_GENERAL = STRUCTURE

```
                        identification :              STRING;
                        cad/cam_system_coming_from :
                                      REFERENCE(CAD/CAM_SYSTEM);
                        creating :                    CAUSAL;
                        reception :                   TIME_STAMP;
                        adm_data_changing :           CAUSAL;
                        model_data_changing :         CAUSAL;
                        host_coming_from :            HOST;
                        variant :                     STRING;
                        version :                     STRING;
                        release_of_the_model :        RELEASE;
                        in_change :                   LOGICAL;
                        valid_until :                 ANY (DATE);
                        remarks :                     STRING;
            END;
```

The entity ADM_DATA_GENERAL represents the general part of the administrative data, containing attributes of general validity necessary for all types of models.

Attribute name	Meaning and values
identification	Internal identification of the set of administrative data, a unique string
cad/cam_system_coming_from	The CAD/CAM-system the model comes from
creating	Date and time when the user created the model and userid of the user who created the model
reception	Date and time when the model and its administrative data were received by (input to) the database
adm_data_changing	Date and time when the user changed the administrative data and userid of the user who changed the administrative data
model_data_changing	Date and time when the user changed either the structure of an assembly (say by adding a new element) or the geometry in a B_Rep that doesn't cause a new version of the geometry (correcting a mistake, say)
host_coming_from	Host the model comes from
variant	There can exist several variants of a special model
version	Each variant of a model can exist in several versions
release_of_the_model	The valid release (status and date) of the model
in_change	Flag to show, if the model itself is currently being changed by a user (dealing with the original, synchronizing of work)

valid_until | The date up to which the model is valid; having received this date the model will be automatically deleted to reduce data abundance

remarks | Possibility for the user to make unstructured remarks concerning the model

Identification is an internal identification number to identify sets of administrative data. The identification will be generated during the INSERT of the set of administrative data and consists of the actual date and, time and, in the case of a distributed system, a code of the host where the set is stored.

2.3.6 Specific Administrative Data for Part or Tool Geometry

```
ENTITY ADM_DATA_GEOMETRY = STRUCTURE
                                title :                          STRING;
                                geometry_format :                ENUM (VDAFS, IGES,
                                                                 SET, NATIVE,
                                                                 CAD*I_NEUTRAL_FORMAT)
                                dimension :                      DIM;

                                company_specific_data =
                                STRUCTURE
                                        part_code :                       STRING;
                                        model_code :                      STRING;
                                        unified_parts_grouping :          STRING;
                                        model_change_level :              STRING;
                                        country_spec_package :            STRING;
                                        optional_equipment_code :         STRING;
                                        engineering_status :              ENUM (DELAY, OK,
                                                                          EXECUTED);
                                        unit_of_length_measurement : ENUM (MM,
                                                                          INCH);
                                        place_of_deposit :                STRING;
                                        drawing_number :                  INTEGER;
                                        drawing_change_index :            STRING;
                                        sheet_number :                    INTEGER;
                                END;
                        END;
```

The entity ADM_DATA_GEOMETRY contains special administrative data necessary to describe sufficiently a part or tool geometry. They must be regarded in addition to the common valid administrative data, look at entity ADMINISTRATIVE_DATA.

Administrative data specific to geometry are different for all kinds of companies, therefore this entity must be regarded variable. The first set of attributes has a general meaning, the following attributes, the company_specific_data, can be defined by each company itself. Here we show an example for the extension valid for the car industry.

Attribute name	Meaning and values
title	The name of the part or the tool
geometry_format	Values: VDAFS, IGES, SET, NATIVE, neutral file format (CAD*I)
dimension	The dimension of the engineering, which can be two-, two-and-half or three- dimensional
part_code	Id-number of a part (an engineering part) or tool
model_code	Classification of the product spectrum of a company into special categories
unified_parts_grouping	A summary of part geometries and assemblies in a logical way; hierarchical system of keys divided into main keys and sub-keys
model_change_level	The part geometry gets a new index if a defined number of changes has been made
country_spec_package	Name of the country in which this special design of the part geometry is valid
optional_equipment_code	Identification of non-standard parts
engineering_status	Status to show the punctuality - delay - just in time - executed
unit_of_length_measurement	mm, inch
place_of_deposit	In the sense of a drawing administration
drawing_number	Identification of the drawing in which the part/tool geometry is contained
drawing_change_index	A drawing receives a new index, if the drawing itself has been changed, but not the drawn part
sheet_number	If a drawing consists of several sheets

2.3.7   Example of a Set of Administrative Data

The entity ADMINISTRATIVE_DATA describing an assembly which consists for example of two elements, containing part geometry as closed models, looks like:

ENTITY ADMINISTRATIVE_DATA =		possible values:
STRUCTURE		
identification		01870513125503
c/c_system_coming_from	name	CATIA
	rel ease	2.0
creating	date	870510
	time	071313
	userid	CADI 012
reception	date	870513
	time	125503
adm_data_changing	date	870622
	time	080310
	userid	CADI 007
model_data_changing	date	870623
	time	083317
	userid	CADI007
host_coming_from	type	MICROVAX
	operating_system	VMS
	operating_system_rel	4.6
variant		WITH SQUARE BORE-HOLES
version		1.0
release_of_the_model	status	EXPERIMENT RELEASE
	status_date	870510
in_change		NO
valid_until		--
remarks		STILL NO EXPERIENCES WITH THIS VARIANT
title		BUFFER BAR
geometry_format		CAD*I_NEUTRAL_FORMAT
dimension		D3
part_code		4711.3/000
model_code		E79
unified_parts_grouping		10.33
model_change_level		A
country_spec package		USA
optional_equipment_code		--
engineering_status		DELAY
unit_of_length_measurement		MM
place_of_deposit		--
drawing_number		739598641
drawing_change_index		0
sheet_number		1
END;		

2.4 Geometric Models

In this section we describe the geometric model entities as used in the CAD*I data base. The definitions are taken from the original CAD*I schema and simplified for the scope of implementation.

2.4.1 Geometric_Model

```
ENTITY GEOMETRIC_MODEL =        CLASS (    SOLID_MODEL,
                                           CLOSED_MODEL,
                                           INSTANCE(GEOMETRIC_MODEL) );
```

The entity GEOMETRIC_MODEL is a class representing entities according to their geometric modeling techniques. For the CAD*I data base we will only use the SOLID_MODEL and the technique of CLOSED_MODEL.

2.4.1.1 Closed_Model

ENTITY CLOSED_MODEL = sequence of binary dates.

The entity CLOSED_MODEL represents geometry as a sequence of binary dates, without any structure or interpretation.

2.4.1.2 Solid_Model

```
ENTITY SOLID_MODEL = CLASS(      B_REP,
                                 INSTANCE(SOLID_MODEL) );
```

A solid may be a B_REP. In WG4 we will only use the B_REP-MODEL.

2.4.2 Boundary Representations

```
ENTITY B_REP =              STRUCTURE
                                SCOPE;
                                    B_REP_SCOPE_ENTITY;
                                END_SCOPE;
                                result:     LIST OF REF_ONLY(SHELL);
                            END;

ENTITY B_REP_SCOPE_ENTITY =      CLASS (    POINT, DIRECTION(D3),
                                            LINE, PLANAR_SURFACE,
                                            VERTEX,
                                            EDGE,
                                            EDGE_LOOP,
                                            FACE,
                                            SHELL);
```

A B_REP is an entity that has a scope. A B_REP may have a material property associated with it. The B_REP is a 'self-contained' entity in the sense that no entity in the B_REP may refer to an entity outside the scope of the B_REP. All referenced entities must be within the B_REP scope itself. The scope of a B_REP contains both topological and geometrical entities. Geometry is represented by lists of the entities POINT, DIRECTION, LINE and PLANAR_SURFACE which are referenced by the topological entities defined subsequently. Topology is represented by lists of the entities VERTEX, EDGE, EDGE_LOOP, FACE and SHELL in that order so that no entity is referenced before it is defined.

The B_REP data structure comprises both topological and geometrical entities. Though checks for consistency of a model are assumed to be the responsibility of the receiving CAD system the following constraints apply:

1. The object must be a two-manifold.

2. The topological structure has to fulfill the Euler formula

$$V - E + 2F - L - 2(S - G) = 0$$

V = number of vertices
E = number of edges
F = number of faces
L = number of loops
S = number of shells
G = genus of the solid object, i.e., number of through holes or number of "handles"
    or "tunnels"

This formula still allows for different valid topological structures for the same object. The objectives for creating a topological structure for a given object on the basis of the Euler formula are, therefore,

-   to avoid artifact edges

-   to use the minimal topological structure
    (the topological structures for cylinder, cone, sphere and torus that fulfill these requirements are presented in the WG2-Report.)

3. A B_REP must have one and only one peripheral SHELL

4. A B_REP may have zero or more interior void SHELL entities

5. The result attribute of a B_REP consists of an ordered list of references to SHELL entities. The first entry in that list is a reference to the SHELL that represents the outer boundary of the B_REP.

2.4.2.1 Shell

ENTITY SHELL =   STRUCTURE
                                     shell :   LIST OF REFERENCE (FACE);
                              END;

A shell is a topological entity defined within the scope of a B_REP. It is defined by a list of references to its bounding faces. The order of the list is arbitrary but the faces must be connected and form a continuous surface which divides the three-dimensional space into two distinct regions.

- Every SHELL is either the external boundary of a B_REP or an internal void within a B_REP.

- Every SHELL consists of one or more references to FACE entities.

2.4.2.2 Face

```
ENTITY FACE =              STRUCTURE
                                  surface_ref :  REF_ANY(PLANAR_SURFACE);
                                  orientation :  LOGICAL;
                                  loop :         LIST OF REF_ONLY(EDGE_LOOP);
                           END;
```

A FACE is a topological entity defined within the scope of a B_REP. It is defined in terms of a set of bounding LOOP entities, the underlying surface geometry, and an orientation.

1. A FACE must belong to one and only one shell.

2. A FACE must have as its geometry one and only one FACE_SURFACE.

3. The ORIENTATION flag of the FACE indicates whether the normal direction of the FACE agrees with the normal direction of the underlying surface geometry. ORIENTATION is .T. if the normal directions agree, otherwise it is .F. Intuitively this indicates whether the solid-to-void direction at any point on the associated surface agrees (.T.) or not (.F.) with the implied normal to the surface at that point. In the case of a PLANAR_SURFACE, ORIENTATION is .T. if the normal points from solid to void.

4. A FACE must be bound by one or more loops, understood in the sense that for an EDGE_LOOP the associated surface is bounded by the closed contours built by the edge curves associated with its edges.

5. Every EDGE_LOOP bounding a FACE must be oriented such that the following inequality holds:

    (FACE_ORIENTATION * LOOP_ORIENTATION) ** FACE_SIDE > 0

    - * denotes the vector (cross) product.
    - ** denotes the scalar product of vectors.

    At a given point on an edge curve of the LOOP the vectors are defined as follows:

FACE_ORIENTATION :	pointing from solid to void
LOOP_ORIENTATION :	pointing in the direction of the edge curve
FACE_SIDE:	tangent to the surface associated with the face and pointing from the edge-curve into the face.

Intuitively this means that an observer standing on the surface on the void side and walking along the LOOP has the finite part of the surface on his left.

2.4.2.3 Edge_loop

```
ENTITY EDGE_LOOP = STRUCTURE
                    loop :      LIST OF STRUCTURE
                                    edge_ref : REFERENCE(EDGE);
                                    edge_orient : LOGICAL;
                                END;
                    END;
```

An EDGE_LOOP is a topological entity defined within the scope of a B_REP.

1. A EDGE_LOOP must be part or all of the boundary of one and only one FACE.

2. Every EDGE_LOOP consists of an ordered list of unique references to one or more EDGE entities. The list of edges should define a continuous path in an anti-clockwise sense when viewed from void to solid. Adjacent EDGE entities in the EDGE_LOOP have a common VERTEX as do the first and last EDGE entities.

3. Every reference to an EDGE in an EDGE_LOOP has an ORIENTATION flag associated with it, to indicate whether the sense of the EDGE in the LOOP coincides with the sense of the LOOP. The ORIENTATION flag is .T. if the senses coincide, otherwise it is F.

2.4.2.4 Edge

```
ENTITY EDGE =               STRUCTURE
                                curve :             REFERENCE(LINE);
                                start_vertex :      REFERENCE(VERTEX);
                                end_vertex :        REFERENCE(VERTEX);
                            END;
```

An EDGE is a topological entity defined within the scope of a B_REP. It is defined in terms of its bounding vertices and its underlying LINE geometry.

1. Every EDGE has one and only one start VERTEX.

2. Every EDGE has one and only one end VERTEX.

3. Every EDGE refers to one and only one underlying LINE.

4. Every EDGE is defined by the sequence of its start VERTEX and end VERTEX such that it coincides with the sense of the underlying LINE.

5. Every EDGE whose start VERTEX and end VERTEX refer to the same point has a closed curve as underlying LINE. In that case the sense of the EDGE is defined arbitrarily.

6. Every EDGE occurs exactly twice, once in each direction, in the lists of EDGE entities referenced by LOOP entities.

The geometry associated with the EDGE is defined as a curve that is common to the surfaces which are associated with all FACES adjacent to that EDGE. In this cases, it will be the intersection of the two surfaces.

2.4.2.5 Vertex

```
ENTITY VERTEX =              STRUCTURE
                                   point_ref :    REFERENCE(POINT);
                             END;
```

A vertex is a topological entity defined within the scope of a B_REP.

1. Every VERTEX consists of a reference to one and only one point

2. A VERTEX may be referenced from one or more EDGE entities

3. The PLANAR_SURFACE geometry in a B_REP has the highest priority in defining the geometry of a B_REP.

This means that the POINT geometry associated with the VERTEX is considered to be the best approximation available of the point that is common to all surfaces associated with the faces that meet at that particular VERTEX.

2.4.3 Geometry Representation

2.4.3.1 Planar_surface

```
ENTITY PLANAR_SURFACE = STRUCTURE
                                   point :              ANY(POINT);
                                   normal :             ANY(DIRECTION);
                                   reference_point :    ANY(POINT) ;
                             END;
```

1. The surface is parameterized.

2. The point is a point in the plane.

3. The reference point should be approximately in the plane and is used to define an axis in the plane for parameterisation.

If the reference point is not given an arbitrary point satisfying the above condition may be used.

2.4.3.2 Line

```
ENTITY LINE =                STRUCTURE
                                   point:        ANY(POINT);
                                   direction :   ANY(DIRECTION);
                             END;
```

The LINE is a geometrical entity defined as an elementary straight line going through the specified point with the specified direction. The point may be defined locally or be represented by a reference to a point entity. The same applies to the direction. The line, like all elementary curves, has an implied curve parameter and is parameterized. If several lines refer the same point they will always go through the same point even when that point is moved interactively. Similarly, if several lines refer to the same direction entity they will always be parallel.

2.4.3.3 Point

ENTITY POINT=   STRUCTURE
                                   x :   REAL;
                                   y :   REAL;
                                   z :   REAL;
                       END;

A POINT identifies a location in three-dimensional Cartesian space by giving its coordinate values x, y, and z.

2.4.3.4 Direction

ENTITY DIRECTION =   STRUCTURE
                                   x :   REAL;
                                   y :   REAL;
                                   z :   REAL;
                       END;

The DIRECTION entity defines the x-, y-, and z-components of a direction vector. The direction vector in the reference schema in normalized.

2.5 Access Rights

Protection of data against inadmissable access is today a very important subject, not only of personal data but also of new developments, even constructions made using CAD/CAM systems. Therefore a neutral database interface is also required to be supplied with aids and instruments to control access to the database.

We have to distinguish between the following kinds of access rights:

Access control

To be able to use the database via the neutral interface the user/application program (in the following simply called "user") must possess an authorization, which can be regulated by special userids and individual passwords. With these userids it is possible to define user classes, such as

-   (normal) user

-   system manager

For each class the available functions are defined. For example, only a system manager is allowed to (or has to) generate authority groups and to insert or delete them in the database.

Access authority

Each set of administrative data and thus each geometry can only be accessed by authorized users. The philosophy in the world of CAD/CAM is to protect each geometry against inadmissable access, either read or write or delete. This is a very strong requirement of CAD/CAM users.

A user must be able to define very individually the access rights for his geometries. Therefore it is necessary to assign to each geometry a set of access rights by the user during the insert into the database.

Protection of single fields (of administrative data)

Some fields in the set of administrative data are not allowed to be changed by the user, not even if he possesses the access right "WRITE" for this set of administrative data. In general these are fields whose values are generated by the system, such as IDENTIFIER, ADM_DATA_CHANGING and so on. At the level of fields there exists not only update protection but also insert protection and, concerning the dialogue, display protection (e.g. IDENTIFIER)

The moment of access control

At the beginning of each transaction (SELECT, INSERT, UPDATE, DELETE of complex objects (see Sect. 3.2) the system has to examine whether the calling user is allowed to put through this transaction or not. For the single functions inside a transaction such as "INSERT ACCESS RIGHT" the check is not necessary, because it has already been done in connection with the corresponding set of administrative data.

Restriction of access rights

Dependent on the values of certain administrative data, existing access rights can be restricted. For example,

STATUS = 'official'

has the consequence that the geometry (and its administrative data) is not allowed to be deleted by anyone, not even by users who possess the access right DELETE for this geometry.

2.5.1 Access Authority

Access to a geometry is only possible via the set of administrative data, which means the user always first has to select the desired set of administrative data (perhaps in two steps, if the selection criteria deliver more than one set). Only then can the user access the geometry (and also the declared access rights) belonging to this set of administrative data. In this manner administrative data possess a control function (see Sect. 3.2).

```
ENTITY AUTHORITY =        STRUCTURE
                              who_has_access: REFERENCE (CLASS (USER, GROUP));
                              which_access : ENUM (READ, WRITE, DELETE);
                          END;
```

AUTHORITY contains information on who possesses which kind of access to which set of administrative data. Each set of administrative data in the database must have one but can also have several authorized groups assigned.

Attribute name	Meaning and values
who_has_access :	Specifies who is allowed to read, to write or to delete the protected model.
which_access :	Specifies which kind of access the authorized person possesses.

2.5.1.1 Access Authority Concerning Closed Models

Each geometry has a set of administrative data and a number of access rights saying which users are allowed to READ, which are allowed to WRITE (and thereby implicit to READ) and which are allowed to DELETE (and thereby implicit to READ and WRITE). These access rights relate to the corresponding set of administrative data.

An access right has the following structure:

GROUP :   group of one or more users and/or groups for which the access right is declared

TYPE_OF_ACCESS :   type of access declared for the group; this may be READ, WRITE or DELETE

An example based on Fig. 2.22 follows:

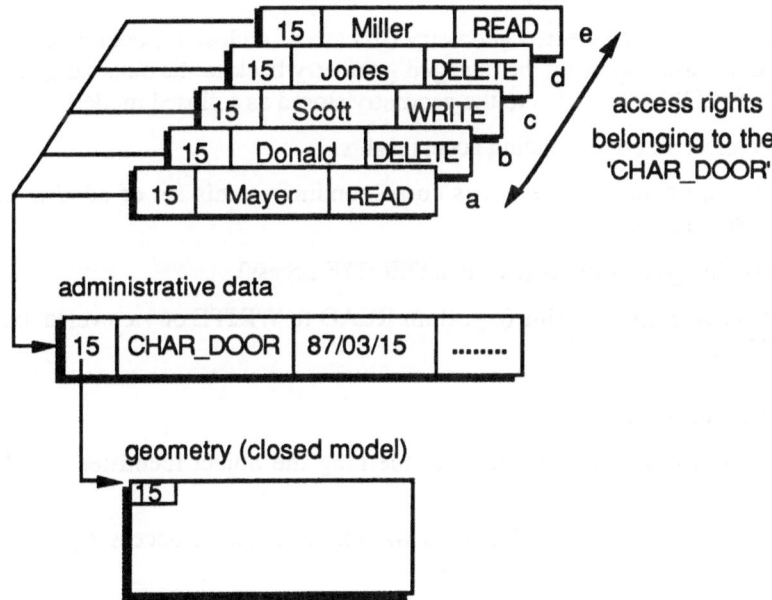

Fig. 2.22 Example for the use of access authorities

Access right READ (a)

Mayer possesses the access right READ concerning the complex object identified by "15", which means

- Mayer can read this set of administrative data, which means she can see the values if this set fullfills the selection criteria. Another user, maybe Mr. Black, will not even know about the existence of this set; nevertheless he will use the same selection criteria.

- Mayer can select (read) all corresponding access rights (a -e). Thereby she knows whom to ask to be given the access right 'DELETE' or 'WRITE' for this set of administrative data (look at access right c).

With this access right READ the user can only inform her/himself via the administrative data (which describe the geometry) about which geometries already exist. This is very important and necessary information for the technical designer to keep the construction (depending on other constructions) up to date.

Access right WRITE (c)

Scott possesses the access right WRITE concerning the object identified by "15", which means

- Scott is able to select (read) the set of administrative data.

- He is allowed to change the values in this set of administrative data except in the update protected fields.

- He is able to delete any single fields of the set of administrative data by updating with blanks except the update protected fields.

- He is able to select the corresponding geometry (only for application programs which are able to work it up, not for dialogue!).

- He is able to change the selected geometry (in case of a closed model he can only do so outside the data base and bring the changed geometry back to the data base, which means he possesses WRITE access also to the geometry stored as a closed model).

- He is able to select all corresponding access rights.

- He is able to insert new access rights corresponding to this set of administrative data (except a delete access).

- He is able to delete access rights (except a DELETE access).

- He is able to update access rights (e.g. from READ to WRITE or vice versa, but not from READ/WRITE to DELETE).

Access right DELETE (d):

Jones possesses the access right DELETE concerning the object identified by "15", which means

- Jones possesses all rights mentioned in the explanation of access rights READ and WRITE.

- In addition he is allowed to delete selected set of administrative data by which the corresponding geometry and all its access rights will automatically be deleted (this provides a guarantee of consistency!).

- He is allowed to insert and to delete the access right DELETE (or to update it).

2.5.1.2 Access Authority Concerning B_Rep

In the following, "B_Rep" denotes a geometric model divided into its points, directions, lines, planar-surfaces and topology (and not stored as a closed model).

Each B_Rep must also have a set of administrative data so that it can be selected by other (describing) values than its exact name. Therefore the access to a B_Rep should be possible only via its administrative data, where the access rights are controlled.

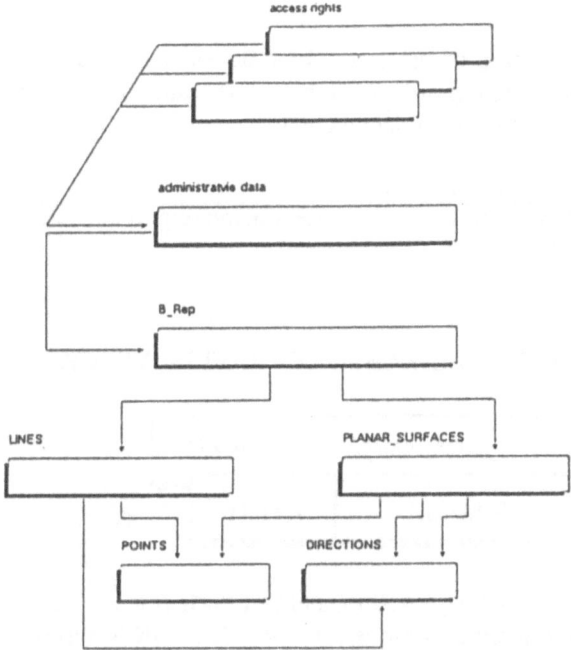

Fig. 2.23 Access authority concerning B_Rep

It does not make sense to check the access rights for, each update of a single point or direction, for example. Rather, such functions have to be collected into well defined transactions (UPDATE B_REP, INSERT B_REP,...) so that at the beginning of each transaction the special access must be checked.

For example, a transaction UPDATE B_REP can be any selection of the following functions:

INSERT POINT, LINE, DIRECTION,...

UPDATE POINT, LINE, DIRECTION,...

DELETE POINT, LINE, DIRECTION,...

because inserting a line, say, into a B_Rep means an updating of this B_Rep. To be able to insert e.g. a line into a B_Rep the user needs access right WRITE for this B_Rep.

2.5.2 Authority Groups

To avoid writing the same list of persons (as authorized persons for administrative data) again and again, it should be possible to collect them to a group which can be addressed by a unique name (unique in the world or unique and valid only in an assembly).

ENTITY GROUP =   STRUCTURE

                         groups :   LIST OF REFERENCE (CLASS (USER,

                                           GROUP));

             END;

The entity GROUP represents a collection of users to which access rights are assigned. A GROUP can be one single user or one single group or can be any mixture of users and groups. The definition of a group is recursiv, but may not contain circles!

Attribute name	Meaning and values
groups:	List of all groups and users belonging to the defining group

Example

group CAD*I_WG4 consists of: FELDMEIER,GRANDL, RAFLIK, WEICK.

access rights:

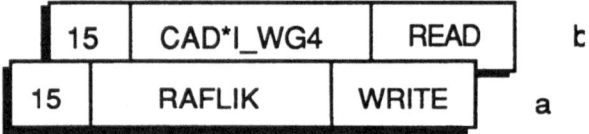

WEICK, GRANDL, RAFLIK and FELDMEIER are allowed to read the set of administrative data; only RAFLIK is allowed to update them. The access right (a) in this case overlays the implicit contained READ access of RAFLIK in (b).

access rights:

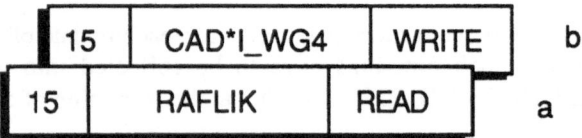

The access right READ given to RAFLIK (a) in this case is senseless, because he is a member of the group CAD*I Database with access right WRITE (b) and this contains the access right READ.

A group can also contain other groups beside users which allows a hierarchical structure to be built up to reflect, say, a department hierarchy. An example is shown in Fig. 2.24.

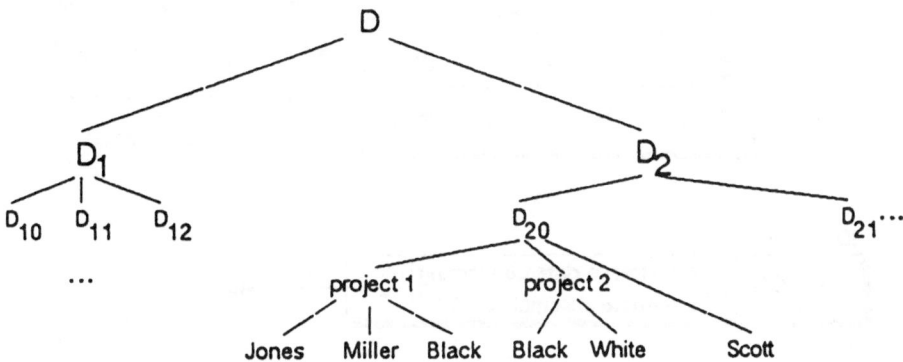

Fig. 2.24 Example of a hierarchical group structure

2.5.3 Example

In the example (fig. 2.25) the access rights a, b, c and d concern the whole assembly, which means: Scott (a) and all members of the group CAD*I_WG4 (c) possess the access right READ for the assembly "CAR". They are allowed to read every geometric model ("DOOR", "MOTOR") of the assembly.

Jones possesses the access right WRITE (d) for the assembly "CAR", which means

☞ he possesses the access right WRITE for each geometric model ("DOOR", "MOTOR") of the assembly with all the rights and possibilities mentioned in the explanation of access right WRITE

☞ he is allowed to update /delete/insert values of the administrative data of the assembly

☞ he is allowed to update/delete/insert access rights at the assembly level

☞ he is allowed to update the assembly, that is, to insert/delete geometric models in the assembly.

Miller possesses the access right DELETE (b) for the assembly "CAR", which means he implicitly possesses READ and WRITE access and in addition he is allowed to delete the whole assembly with all geometric models with all corresponding sets of administrative data and access rights.

But there are also access rights necessary at the geometric model level, which supplement or restrict the access rights at the assembly level. In the example, to the geometric model "DOOR" there belong the access rights (a, b, c, d) implicitly and (e, f) explicitly, whereby at this level SCOTT possesses two different access rights a = READ and f = DELETE, for the geometric model "DOOR". In this case we have to introduce one of the following regulations:

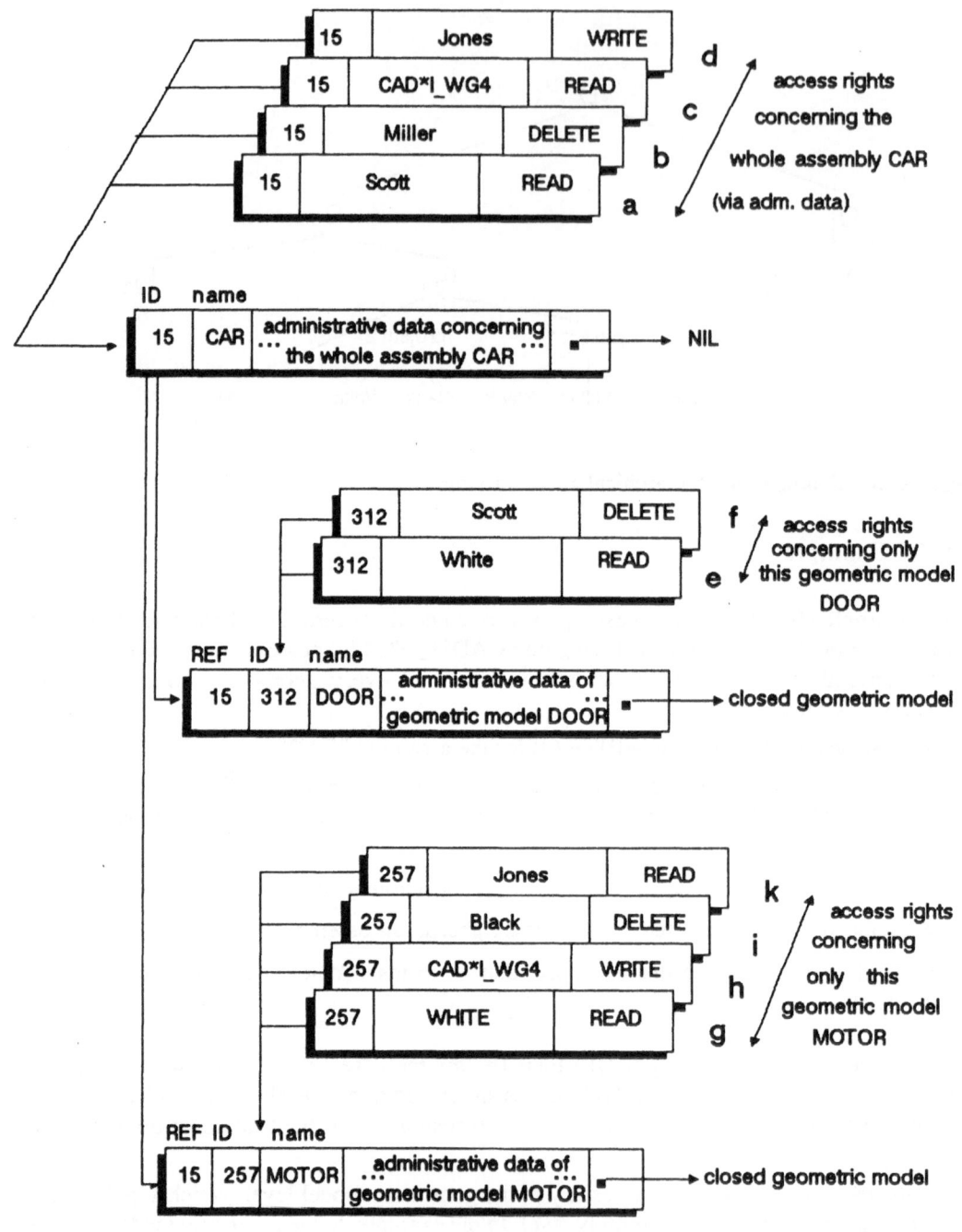

Fig. 2.25 Example for use of access authorities for assemblies

Alternative 1

The access right with the highest priority is always the valid one, with the priority:

1 DELETE

2 WRITE

3 READ

In the example this means Scott is allowed to read all geometric models of the assembly by access right (a) and especially to delete the geometric model "DOOR" by access right (f).

Alternative 2

The access rights belonging explicitly to the geometric model of the assembly are always dominant and may override colliding access rights at the assembly level.

In the example this means Jones is allowed to write all geometric models of the assembly by access right (d), except the geometric model "MOTOR" which he is only allowed to read by access right (k).

3   Specification of the Neutral Interface Routines

3.1   General Rules for Building Routines

In this chapter the general rules and ideas for defining neutral interface routines are presented, especially regarding how to generate the names of the routines, how to generate names and the order of the used parameters, and how to handle scopes, names and references.

Some other important facts should be noticed:

☞   The routines follow the FORTRAN-77 standard

☞   The routines are only procedures and not functions

☞   All arrays used are only one-dimensional to facilitate links to other program languages.

3.1.1   How to Generate the Names of the Routines

The naming of the routines conforms to the following rules:

☞   Every routine has a name of six characters like 'NIBFEN'.

☞   The first two characters 'NI' of the routine name express that this is a Neutral Interface routine of the CAD*I Project.

☞   The next two characters 'BF' stands for the basic function of the routine (see also Table 3.1):

a)   SL :   **SeL**ect routine

b)   IN :   **IN**sert routine

c)   UP :   **UP**date routine

d)   DL :   **DeL**ete routine

e)   LI :   **LI**st routine

f)   EC :   **EX**clude routine

g)   IC :   **InC**lude routine

h)   OP :   **OP**en routine

i)   CL :   **CL**ose routine

j)   EN :   **EN**d Selection routine

k)   ET :   **E**nd of **T**ransaction routine

☞ Characters 5 and 6 ('EN') indicate the type of entity handled in the routine. The two characters used for indicating this are given in Table 3.1 in column EN.

Examples:

The name of a procedure to insert a point into the database is: NIINPT

The name of a procedure to update an edge in the database is: NIUPED

The name of a procedure to delete a set of administrative data in the database is: NIDLAD

Every routine has a unique integer-number identification which is used for example in error-handling. This number can be generated from Table 3.1 by adding the value of the column NR and the value of the basic function.

Examples:

Routine NIINPT has the number	$100 + 2 = 102$
Routine NIUPED has the number	$600 + 3 = 603$
Routine NIDLAD has the number	$2000 + 4 = 2004$

3.1.2 How to Generate the Names of the Parameters

The naming of the parameters conforms to the following rules:

☞ Every parameter has a name of six characters like XYZPAR

☞ The first character 'X' stands for the information of the data type (according to the FORTRAN-77-Standard):

a) C :   Character

b) I :   Integer

c) L :   Logical

d) R :   Real

e) S :   String

☞ The second character 'Y' indicates whether we have a single parameter or an array of parameters:

a) A :   Array of parameters

b) S :   Single parameter

☞ The third character 'Z' indicates the read/write- access of the routine to the parameters:

a) F :   read/write (Free) access

b) R :   Read-only access

c) W :   Write-only access

☞   Characters 4 to 6 ('PAR') are free for selection.

Examples:

LSRABC: This is a single parameter of data-type LOGICAL which is only read by the
            routine. This ensures that the value of LSRABC is not changed by the routine.

IAWERR: This is a parameter-array of data-type INTEGER which is only written by the
            routine. The value of IAWERR before calling the routine has no influence on the
            routine, whereas IAWERR will be changed by the routine.

ISFNAM: This is a single parameter of data-type INTEGER that can be used as an input
            parameter as well as an output parameter.

3.1.3 Naming and Referencing of Entities

As agreed in the CAD*I project, the entities are defined by positive integer numbers. This fact
is kept in the neutral interface routines. Another point is the usage of the 'REF_ANY' attributes
- entities which is solved as follows:

☞   Entities in the database have positive integer names.

☞   Attributes in the database are treated as entities with negative integer names as far as
    naming and storing is concerned.

    Note:

    The routines must guarantee that no structured attribute is used in more than one place and
    that upon complementation of a scoped entity no structured attribute remains unused. This
    is achieved in the following way:

    a) The system maintains a table (NI_UNUSED_ATTRIBUTE) of defined but unused
       attributes.

    b) Whenever an attribute is inserted ("entity with negative number"), its name is entered
       into this table.

    c) Whenever a negative entity number is used for defining an attribute of another entity, it
       will be searched for in this table. If found, it will be removed from this table. If not
       found, an error message will be issued, even if the data base contains that "entity".

    d) Whenever a "CLOSE" routine is called, the table will be searched for "entities". If any
       remain, an error message will be issued and the table will be cleared.

☞   Instead of the REF_ANY mechanism we have only one mechanism, namely reference by a
    (positive or negative) integer number.

☞   To identify an entity or an attribute in the database by its name, means to build an identifier
    out of its name, its scope-number and the membership of a definite table in the relational
    database system.

Therefore we can guarantee that every entry in the database can be exactly retrieved.

3.1.4 Scope Mechanism

There are some entities with a scope. In the routines the scope mechanism is hidden from the application by its purely internal usage. If an entity with a scope is inserted into the database, then a new and unique scope number is generated by the system. All the following routines react in respect to this scope number.

3.1.5 Error Handling

Every routine returns an error code in the parameter IAWERR. IAWERR is an integer-array of three values, which are returned by the routines with the following meaning:

1. IAWERR(1) :   kind of error

2. IAWERR(2) :   number of the called routine

3. IAWERR(3) :   number of the error/message

If the routine detects an irreparable error, the routine will not make any transaction in the database, which means the content of the database will be the same as before the call of this routine. This is supported by returning a meaningful content of the parameter IAWERR to the calling module.

If any module calls any NI-routine, it can check by this parameter, whether the procedure was completed correctly. If an error is detected, the module can display the error and the list of the calling modules (tracing) to the user by calling a special error-handling routine (NISLER).

If there is an error that cannot be handled by the neutral interface routines, a database error handler gives his specific message to the calling module.

3.1.6 General Form of the Specified Routines

Every routine is defined in the same way:

☞ **Procedure**
   In this part the name of the procedure and its arguments are defined in FORTRAN notation.

☞ **Parameters**
   In this section the parameters are defined in respect to their data-types.

☞ **Input**
   In this section the parameters containing input information for the routine are described.

☞ **Output**
   In this section the parameters containing output information for the routine are described.

☞ **Function**
   Here the main function of the routine is described.

☞ **Remark**

The remark section is used for mentioning necessary checks done by the routine and noticing specialities of the routine.

☞ **Examples**

Here one or more examples are presented to show the usage of the routine. The parameters used in the examples are assumed to be defined correctly in the calling program according to the parameters section of the specified routine.

☞ **Possible Error Messages**

In the last section of the specification the main messages and their error status is described. This section may grow during the implementation phases.

3.1.7 Basic Functions

Table 3.1 presents an overview of the specified routines and shows the basic functionalities with their code:

SL = Select	Code 1
IN = Insert	Code 2
UP = Update	Code 3
DL = Delete	Code 4
LI = List	Code 5
OP = Open	Code 6
CL = Close	Code 7
IC = Include	Code 8
EC = Exclude	Code 9
EN = End of Selection	(Routine applied to FUNCTION)
ET = End of Transaction	(Routine applied to FUNCTION)
CT = Cancel Transaction	(Routine applied to FUNCTION)

The following table also lists the entities used with their full name (column ENTITY), their sign (column EN) and their related number (column NR). Additionally the error routine is mentioned:

'+'   means this routine is specified,
'-'   means this routine will not be specified,
' '   means this routine may be specified at a later date.

Note:

The DELete-procedures in the present specification makes no consistency check on the model level.

NR	EN	ENTITY	SL 1	IN 2	UP 3	DL 4	LI 5	OP 6	CL 7	IC 8	EC 9
100	PT	POINT	+	+	+	+	+	-	-	-	-
200	DI	DIRECTION	+	+	+	+	+	-	-	-	-
300	LI	LINE	+	+	+	+		-	-	-	-
400	PS	PLANAR-SF	+	+	+	+		-	-	-	-
500	VE	VERTEX	+	+	+	+		-	-	-	-
600	ED	EDGE	+	+	+	+		-	-	-	-[a]
700	LP	LOOP	+	+	+	+		-	-	-	-
800	FA	FACE	+	+	+	+		-	-	-	-
900	SH	SHELL	+	+	+	+		-	-	-	-
1000	BR	BREP	+	+	+	+	+	+	+	-	-
2000	AD	ADM_DATA	+	+	+	+	+	-	-	+	+
2100	AC	ACC_RIGHTS	+	+	+	+	+	-	-	-	-
2200	CM	CLO_MODEL	+	+	-	+	-	-	-	-	-
3000	US	GROUP (DBA)	+	+	+	+					
3100	TR	GROUP TREE		+		+					
9800	WO	WORLD	-	-	-	-	-	+	+	-	-
9900	ER	ERROR	+	-	-	-	-	-	-	-	-

Table 3.1: Basic functionality of the geometry routines

3.2 Transaction Concept

In database management systems the integrity of data and the security of data are connected with the concept of a transaction.

In [DATE/81] a transaction is declared as

" a unit of work that is atomic from the view of enterprise"

where a unit is said to be atomic when it is a single action that cannot be divided or interrupted. Within a transaction it is guaranteed (by the ACID principle) that for a sequence of operations:

☞ **Atomicity**
the operations cannot be interrupted

☞ **Consistency**
either all operations or none will be performed

---

[a] In this proposal there is no difference between a field that has no value and a field to which blanks are assigned. In both cases there will be no insert into the database. In ORACLE thie field is NULL.

☞ **Isolation**
concurrent accesses will be synchronized

☞ **Durability**
the operations are not visible before the end of the transaction and sequentially can only be removed sequentially by further transactions.

Before the end of a transaction the integrity of data must be guaranteed, which allows only updates which are "rich in meaning" and "admissible".

3.2.1 Transaction Logic

The above defined properties of transactions cannot simply be transferred to CAD databases. A transaction concerning a whole part geometry, for example, includes the period from the beginning of the construction up to a well defined, fixed status. The duration of such a transaction can never be fixed by the CAD designer, because it is not possible to roll back the whole transaction (this may be the whole construction) in case of an error appearing.

The conventional transaction logic does not allow partly finished results to be made visible during a transaction. But this is a very important demand of CAD designers, whose work is dependent on other works or constructions. The visibility of partly finished results can be realized by introducing versions or different release states. The units of the atomicity must be definable as shorter time periods.

In the CAD*I database it is possible, based on the length of transaction, to make (perhaps inconsistent!) partly finished results deliberately visible to the CAD designers. For that purpose a geometry model is stored in the database as a certain version with the corresponding administrative data and is thereby visible to all authorized CAD designers.

In this case the logic of transaction is not referenced to the beginning and to the end of a construction, but now controls the consistent administration of such partly finished results by dealing with such complex objects as:

☞ administrative data

☞ access rights

All these must be regarded as one object, a complex unit of single objects, belonging to one another. A complex object is represented outside the database as an abstract data type, which means manipulation of this object is only possible by well defined operations belonging to this complex object. Thus an indiscriminate access to any given structure in the database can be prevented, and that leads in turn to saving of controls and checks on semantic integrity.

3.2.2 Resources for Realizing Transactions

The responsibility for the correctness of the transactions belongs to the application program, not to the database system. However, modern database systems make resources available to define and synchronize transactions.

A transaction is signed by special database operations.

- END OF TRANSACTION   all operations of a transaction, i.e. all operations made by a user since the last consistent state, are fixed and visible.
- CANCEL TRANSACTION   all operations of a transaction are canceled, i.e. the database is rolled back to the last consistent state.

Following these operations the database system is responsible for the necessary recovery. To synchronize concurrent accesses to objects of the database, each object a transaction will need must be locked in advance (otherwise the use of an object by a transaction would not be noticed by other users/transactions). The lock mechanisms of database systems are not uniform. Both operations END OF TRANSACTION and CANCEL TRANSACTION cancel all active locks .

Remark:

Introducing abstract data types such as the locking of access rights is not necessary, because manipulation of access rights is only possible in connection with the corresponding administrative data, and therefore the locking of the administrative data is sufficient to synchronize concurrent accesses to either administrative data or access rights or geometry.

3.2.3 Example of Transactions Concerning the Complex Object "Administrative Data - Access Rights - Geometry"

Example 1:   INSERT a geometry model[1]
*   INSERT administrative data
*   [INSERT access rights]|INSERT generated access right
*   INSERT geometry model data
*   END OF TRANSACTION / CANCEL TRANSACTION

Example 2:   SELECT a geometry model[1]
*   SELECT administrative data[2]
*   SELECT geometry model data
*   SELECT all corresponding access rights

Example 3:   Inform oneself via administrative data (especially online)
*   SELECT administrative data[2]
*   SELECT all corresponding access rights

Example 4:   DELETE a geometry model
*   SELECT administrative data
*   DELETE administrative data
*   DELETE all corresponding access rights
*   DELETE corresponding geometry model data
*   END OF TRANSACTION / CANCEL TRANSACTION

Remark :

A pure update of geometry model data as a closed model in the database is impossible. In this case one has to delete the whole geometry in the database and then to insert the changed geometry or (if the old version is to be kept) to insert the changed geometry as a new version.

Example 5: UPDATE administrative data and access rights
*   SELECT administrative data$^{(2)}$
*   SELECT all corresponding access rights
*   SELECT FOR UPDATE (=LOCK) administrative data
*   UPDATE administrative data
*   UPDATE I INSERT I DELETE access rights
*   END OF TRANSACTION / CANCEL TRANSACTION

Regarding an assembly we have the following transactions:

Example 6:   INSERT an assembly$^{(1)}$
*   INSERT administrative data of assembly level
*   [INSERT all corresponding access rights of assembly level] I
        INSERT generated access right
*   for each element of assembly:
*   INSERT administrative data of element of assembly
        *   [INSERT all corresponding access rights]
        *   IINSERT generated access right
        *   INSERT geometry model data
*   END OF TRANSACTION / CANCEL TRANSACTION

Example 7:   SELECT an assembly$^{(1)}$
*   SELECT administrative data of assembly level$^{(2)}$
*   SELECT all corresponding access rights of assembly level
*   for each assembly element :
        *   SELECT administrative data of element level
        *   SELECT all corresponding access rights of this set of administrative data at
                element level
        *   SELECT corresponding geometry model data

Example 8:   SELECT an element of assembly$^{(1)}$
*   SELECT administrative data of assembly level$^{(2)}$
*   SELECT administrative data of assembly element$^{(2)}$
*   SELECT all corresponding access rights of selected assembly element
*   SELECT corresponding geometry model data

Example 9:   Inform oneself via administrative data concerning assemblies (especially on-line)
*   SELECT administrative data$^{(2)}$
        (this gives all qualified sets of administrative data at assembly level and of the single
        models, but not those at the level of the assembly elements)
*   SELECT all access rights belonging to assembly level
*   SELECT administrative data of all elements of selected assembly$^{(1)}$
*   SELECT all access rights of selected element of selected assembly

Example 10: DELETE an assembly
*   SELECT administrative data$^{(2)}$ of the assembly level
*   DELETE administrative data of assembly level

*   DELETE all access rights belonging to assembly level
*   DELETE administrative data of all assembly elements
*   DELETE all access rights of all assembly elements
*   DELETE geometry model data of each assembly element

Example 11: UPDATE administrative data and access rights concerning assemblies
*   SELECT administrative data$^{(2)}$ of assembly level
*   SELECT FOR UPDATE (=lock) selected set of administrative data
*   SELECT all access rights belonging to this set of administrative data
*   INSERT I UPDATE I DELETE access rights for each element of assembly.
        It is possible to:
        *   UPDATE element of assembly
        *   SELECT administrative data of element of assembly $^{(2)}$
        *   SELECT all access rights belonging to the selected element
        *   INSERT I UPDATE I DELETE access rights
        *   UPDATE administrative data of the selected element
*   UPDATE administrative data of assembly
*   END OF TRANSACTION / CANCEL TRANSACTION

Example 12: UPDATE assembly structure

To update the structure of an assembly means either
*   to insert new elements (a new geometry will be inserted in the database and included
        into the assembly) or
*   to delete an element (an existing assembly element will be excluded and deleted) or
*   to include an element (an already inserted geometry will be included into (= added to)
        the assembly) or
*   to exclude an element (an existing assembly element will be excluded but not deleted; it
        will still exist as a single model in the database).
*   SELECT administrative data$^{(1)}$ of assembly level
*   SELECT FOR UPDATE (=LOCK) selected set of administrative data
*   INSERT element
        *   INSERT administrative data of element of assembly
        *   INSERT all corresponding access rights
        *   INSERT geometry model data
*   DELETE element
        *   SELECT element of the assembly$^{(1)}$
        *   DELETE administrative data of element of assembly
        *   DELETE all access rights belonging to the element
        *   DELETE corresponding geometry model data
*   UPDATE element (i.e. DELETE an element and sequentially INSERT an element)
*   INCLUDE an element already existing as single model
        *   SELECT administrative data (single model)$^{(1)}$
        *   UPDATE administrative data (insert the reference to assembly)
*   EXCLUDE an element
        *   SELECT administrative data (element level)
        *   UPDATE administrative data (delete the reference to assembly)

*   UPDATE administrative data of assembly level (number of elements and date, time of
       change)
*   END OF TRANSACTION / CANCEL TRANSACTION

Explanation:

(1) use only by program (not by dialogue)

(2) if several sets of administrative data qualify, the user/program has to select one

[]   optional

| alternative

In a first step the database interface of a CAD*I database will offer the set of necessary basic
functions. It is the calling system (application programs, dialogue systems) that has the whole
responsibility for observing and checking the logic of transaction, which means that the
observation of the sequence of the called basic routines and the decision to call END OF
TRANSACTION or CANCEL TRANSACTION is situated at this superior level. Semantic
integrity will be checked in the database interface: depending on special return codes and error
messages, this superior level must make decisions.

Basic routines

(1)   INSERT administrative data
(2)   INSERT access right
(3)   INSERT geometry model data
(4)   UPDATE administrative data
(5)   UPDATE access right
(6)   DELETE administrative data (with all corresponding access rights, geometry data)
(7)   DELETE access right
(8)   SELECT administrative data
(9)   SELECT all access rights relating to a special set of administrative data
(10)   SELECT geometry model data
(11)   END OF UPDATE (consistency checks!)
(12)   END OF INSERT (consistency checks!)
(13)   END OF TRANSACTION
(14)   CANCEL TRANSACTION

3.2.4 Semantic Rules: Consistency and Plausibility Checks

In addition to the definition of the transactions, consistency and plausibility checks must be
defined to complete the database design. Generally not all possible values, combinations of
values or change of values are allowed. By semantic rules (restrictions) it is possible to define
which special objects, attributes and relations are allowed and which are forbidden.

This means that a conceptual schema (defined in HDSL) consists of two parts, the basic
structure and the semantic rules. The basic structure describes which types of objects and types
of relations exist, the semantic rules define in which combinations and with which restrictions
relations may appear.

Classification of plausibility checks

To avoid frequent changes of the source code caused by plausibility checks being changed (especially administrative data), the rules are classified and inserted into database tables. For each class a standard method will be developed to check the relevant rules during the database manipulation (= at runtime).

Semantic rules

The following semantic rules must be checked to maintain consistency:

1. If a set of administrative data is one of an assembly, it is not allowed to point to a geometry.

2. If a set of administrative data is not one of an assembly, it must point to a geometry and the geometry must exist in the database

3. Only sets of administrative data of assemblies are allowed to possess subassemblies, elements (sons).

4. Each user/group can possess only one access right for one set of administrative data, either read or write or delete.

5. A group may not have a user as father in the group hierarchy.

6. Each user must be son of at least one group and this group must be a department.

4 Detailed Description of the CAD*I Database Interface

4.1 Functionality of The Global Routines

This chapter describes the functionality of the routines that are independent of the geometric entities.

4.1.1 Functions for Database Access

a)   **Open World (logon database)**

☞   **Procedure:**
NIOPWO (SSRDBN, SSRDBU, SSRDBP, IAWERR)

☞   **Parameters:**
Input:
SSRDBN   name of the database to be logged on
SSRDBU   the identification name of the database user
SSRDBP   the valid password (access control)

Output:
IAWERR   error status

☞   **Function:**
connection and logon to the database and preparing performance optimization

☞   **Example:**
DBNAME = ´DBDEVELOP´
DBUSER = ´CAD*I´
DBPWD = ´WG4_DB´
CALL NIOPWO (DBNAME, DBUSER, DBPWO)

☞   **Error Messages:**

Kind	No.	Message
1	200	NI: database name empty
1	201	NI: database password empty

b)   **Close World**

☞   **Procedure:**
NICLWO (IAWERR)

☞   **Parameters:**
Output:
IAWERR   error status of the routine

☞ **Function:**
logoff current database

Remark:
If there are any unused but inserted structured attributes in the open world, they are deleted from the database and an error is issued.

4.1.2 Functions for Error- Handling

Select error_message

☞ **Procedure:**
NISLER (ISRUNR,IARERR)

☞ **Parameters:**
Input:
ISRUNR   unit number for the output device
IARERR(1) kind of error
IARERR(2) number of the routine
IARERR(3) number of the error/message

Output:
no values returned.

☞ **Function:**
Selects the meaning of error from the ERROR_KIND_TABLE, the name of the procedure from the PROCEDURE_NAME_TABLE, the message from the ERROR_MESSAGE_TABLE and writes the complete information to the output device given by the unit-number ISRUNR. If there are nested calls of the NI-Routines, their tracing is displayed.

☞ **Remark:**
if ISRUNR is 0 , the output is suppressed .

☞ **Example:**
CALL NISLER(7, IARERR)

4.1.3 Functions for Entity-Name Generation

Select name

☞ **Procedure:**
NISLNA (ISWNAM,LSRTYP,ISRTYP,IARERR)

☞ **Parameters**
Input :
LSRTYP   if LSRTYP is true, a positive name is returned,
           if LSRTYP is false, a negative name is returned
ISRTYP   number of the entity, see column NR in Table 3.1

Output :
ISWNAM   free and unique name for the entity of type ISRTYP
IARERR   error-code

☞   **Function**:
Selects a free and unique entity or attribute name using an intrinsic database function.

☞   **Example**:
CALL NISLNA(NAME,.TRUE.,200,IARERR)
would select a free name for a entity of type DIRECTION

4.1.4   End of Selection

☞   **Procedure**:
NIENFC (ISRCUR, IAWERR)

☞   **Parameters**:
Input:
ISRCUR   number of the cursor which references the buffer containing the selected
records and which is now to be set free, because the result set is no
longer of interest

Output:
IAWERR   error status

☞   **Function**:

Sets free the cursor and thereby the buffer for other selections, because the result set is
no longer of interest. If the cursor is 0 then the select buffer, that was used by the last
select function of a geometry or topology element, will be set free.

☞   **Error Messages**:

Kind	No.	Message
1	123	DB: cursor does not exist
1	139	DB: invalid cursor number

☞   **Example**:
A user wants to inform himself about all models being created in the CAD/CAM
system CATIA: this means a selection of all sets of administrative data where
CAD/CAM system = 'CATIA' and the user is allowed to read.

assign the select criterion
      SSRADA = ' ';
      SADCSN = 'CATIA';

select
            CALL NISLAD (SSRADA, CURS1, ISTAT)

fetch first record
            CALL NIREAD (CURS1, 1, SSRADA, ISTAT)
            (now SSRADA contains the first record of the selected sets of administrative data
            if there are any, otherwise a warning appears)

fetch second record
            CALL NIFENT (CURS1, 2, SSRADA, ISTAT)
            (now SSRADA contains the second record of the result set)

fetch one record after the other till the warning
            "NO MORE RECORD SELECTED" appears

give back the cursor
CALL NIENFC (CURS1, ISTAT)

4.1.5 End of Transaction

☞ **Procedure:**
NIETFC (IAWERR)

☞ **Parameters:**
Output:
IAWERR   error status

☞ **Function:**
COMMIT, that makes all manipulations in the database visible and cancels all locks on rows and tables. Before committing, some integrity checks must be done, for example "are there any local sets of administrative data in global sight?"

4.1.6 Cancel Transaction

☞ **Procedure:**
NICTFC (IAWERR)

☞ **Parameters:**
Output:
IAWERR   error status

☞ **Function:**
ROLLBACK, i.e. undo all manipulations in the database since the last end of transaction and cancel all locks of rows and tables

☞ **Error Messages:**

Kind	No.	Message
1	138	DB: database error concerning the ROLLBACK command

4.1.7 Read Buffer

☞ **Procedure:**
NIREAD (ISRCUR, ISRREC, SSWNTS, IAWERR)

☞ **Parameters:**
Input:
ISRCUR   Number of the cursor referencing the buffer which contains the selected records, e.g. sets of administrative data, access rights, etc.
ISRCUR   is output of each select procedure.
ISRREC   Record number of the record to be read from the buffer (the first, the second, etc.). With this parameter the result set can be scrolled up and down or result picked a special out of the result set.

Output:
SSWNTS   String without a structure. It contains the next record of the selected set

which is stored in the buffer referenced by ISRCUR

IAWERR   error status

☞ **Function:**

Fetching the next record of the buffer and assigning it to the variable SSWNTS. With this function the user is able to get a selected set record by record

☞ **Remark:**

Before calling the procedure NIREAD the user has to call a SELECT procedure. If the user do not need the result set any more, he has to call the END OF SELECTION procedure (NIENFC) to set the cursor free (see Fig. 4.1 below).

☞ **Error Messages:**

Kind	No.	Message
1	123	DB: cursor does not exist
2	124	NI: no (more) records selected
1	139	DB: invalid cursor number

☞ Example:
look at procedure NIENFC

4.2 Functionality of Assembly Routines

To reduce the number of parameters, fields belonging to one another are collected in one string, e.g. the string SSFADA contains all administrative data. To insert, say, a set of administrative data the user has to fill the corresponding fields with the desired values. Fields that will not get a value (e.g. the field remarks in the string SSFADA can be empty) must explicitly be blank.

SSFADA   user_id   host   part code

```
  _____  _____  _____  _____   _____  _____
 |      |//////|  |    |////|  | ... |  |///////////|  |      |
 |_____|_____|  |____|____|__|     |__|_____|__|_____|
```

```
 ___
|///|   filled area for insertion
|___|
|   |   the remaining areas must be blank
|___|
```

The necessary instructions are:

SSFADA = ' '   --   the remaining areas will be blank

SADUIC = 'CAD*I012'   --   userid, (SADUIC is a name of the area
                                         SSFADA (76:96), equivalence)

SADHOT = 'MICROVAX' --   host, SADHOT is a name of the area
                                         SSFADA (112:132)

a.s.o.

Then the procedure INSERT ADMINISTRATIVE DATA is called with the parameter string SSFADA.

To update a set of administrative data, say, the user has to select the set of administrative data and to change the value of the desired fields.

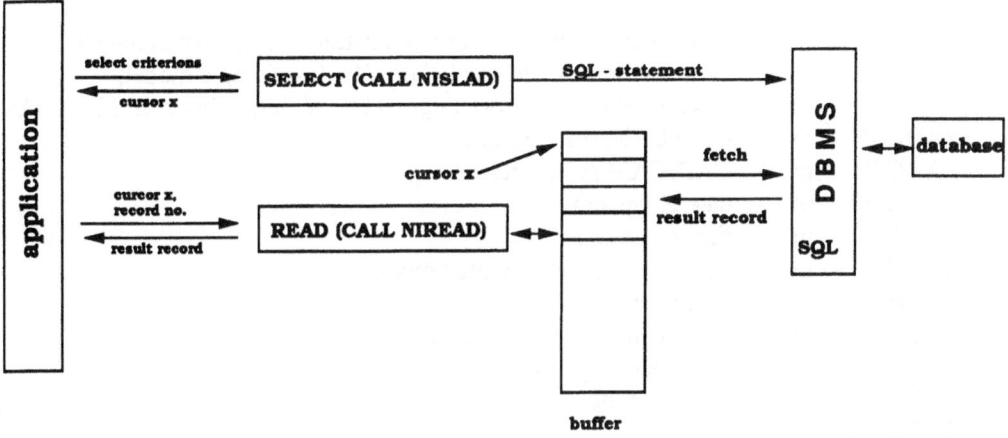

Fig.4.1   Illustration of the principles of selection

Explanation:

The application starts the selection by giving the select criteria to the SELECT-routine NISLAD.

This routine gives back to the application a cursor number, which will reference the result buffer. Then it generates the dynamic SQL-statement and delivers it to the database management system, which processes the selection. The result set will still be held in the database.

Via the cursor (given back by the SELECT-routine) the application is able to fetch one selected record after the other by the READ-routine NIREAD. This routine manages a program buffer as follows. If the application wants to read, say, the first selected record, the READ-routine checks whether this record is already stored in the program buffer or not. If not, it will fetch the record from the database result buffer and put it into the program buffer; this involves database access only when necessary.

By these cursor and buffer techniques it is possible to hold more than one result set in parallel, distinguished by the referencing cursor, and one is able to scroll up and down the selected set of records. The END_OF_SELECTION-routine NIENFC sets the given cursor and the corresponding section of the buffer free.

The mechanism concerning geometry and topology elements, where the selection gives back a list of identifiers of the selected elements, is not useful in the case of administration data, because the identifiers generated by the system do not tell the user anything. He needs the values of the set of administrative data to decide which geometry he wants to work with. The difference is that the selection of geometry and topology elements is mainly made by an application program that knows the identifiers, whereas the selection of administrative data is mainly used by a user who wants to inform himself about existing geometries.

a)   **Insert set of administrative data**

  ☞   **Procedure:**
        NIINAD (SSRADA, IAWERR)

☞   **Parameters:**

Input:

SSRADA   string with the structure of administrative data containing all
                              values to be inserted and informally describing a model.
                              Possible and mandatory (signed by 'm') input fields of SSRADA:

SADCSN   m   CAD/CAM system name
SADCSR   m   CAD/CAM system release
SADADC   m   date of creation
SADATC   m   time of creation
SADUIC   m   userid of user who created model
SADHOT   m   host type
SADHOS   m   host operating system the model comes from
SADHOR   m   host operating system release
SADTOM   m   type of model
SADVAR   variant
SADVER   version
SADMRS   valid release status of the module
SADMRD   release status date of the module
SADMIC   in change
SADMVD   valid until
SADREM   remarks
SADTIT   title
SADGFO   geometry format
SADDIM   dimension
SADLGA   m   local/global administrative data set and the company specific data,
                 e.g. data useful for car industry (whether the data is mandatory or
                 not must also be determined by each company) proposal
SADPCO   m   part code
SADMCO   m   model code
SADUPG   unifid parts grouping
SADMCL   m   model change level
SADCSP   country specification package
SADOEC   optional equipment code
SADEST   m   engineering status
SADULM   unit of length measurement
SADPOD   place of deposit
SADDNR   drawing number
SADDCI   drawing change index
SADSNR   sheet number

All other fields of SSRADA will be generated by this routine

Output:

IAWERR   error status
SSWIDF   the generated identifier for the set of administrative data to
                 allow insertion of access rights.

☞   **Function:**

After successful plausibility checks the automatic values will be generated including a
unique identifier. The so completed set of administrative data will be inserted into the

database. Additionally an access right DELETE for the inserting user will be generated and inserted into the database.

☞ **Error Messages:**

Kind	No.	Message
1	101	NI: creation values missing
1	102	NI: creator must be an authorized user in the system
2	104	NI: the input string is empty / identifier does not exist.
1	206	NI: CAD/CAM system name or CAD/CAM system release is empty
1	207	NI: no unique identifier created
1	208	NI: host type is empty
1	209	NI: host operating system or host operating system release the model comes from is empty
1	210	NI: type of model is empty
1	215	NI: local/global definition empty
1	242	part code must not be blank
1	243	model code must not be blank
1	244	model charge level must not be blank
1	245	engineering status must not be blank

☞ **Example:**

Assign the values to be inserted
    SSRADA = ' '
    SADCSN = 'CATIA'
    SADCSR = 'V2R2'
    SADGFO = 'IGES'
    etc.

insert the set of administrative data
    CALL NIINAD (SSRADA, SSWIDF, ISTAT)

b) **Include a new element (member) into an assembly (owner)**

☞ **Procedure:**
NIICAD (SSRAID, SSREID, IAWERR)

☞ **Parameters:**
Input:

SSRAID	identifier of the set of administrative data of the assembly (owner) into which the element is to be included as a new member
SSREID	identifier of the set of administrative data of the element (member) to be included into the given assembly (owner)

Output:

IAWERR	error status

☞ **Function:**
Generating the connection between owner (this is the set of administrative data of an assembly) and member (this is the set of administrative data of an assembly, B_Rep or closed model). The user needs access right WRITE or DELETE for the assembly that is to be owner or for any owner of this assembly.

☞ **Remark:**
The sets of administrative data of owner and member must already be inserted into the database. A local defined set of administrative data may only be included once. A global defined set of administrative data may be included even in several different assemblies.

☞ **Error Messages:**

Kind	No.	Message
1	107	NI: the element to be included does not exist
1	111	NI: set of administrative data with the given identifier does not exist
1	213	NI: identifier of assembly (owner) is empty
1	214	NI: identifier of data set to be included is empty
1	216	NI: user not allowed to include/update data set
1	217	NI: element to be included already exists in an assembly
1	218	NI: element already included in the assembly

c)   **Exclude an element (member) from an assembly (owner)**

☞ **Procedure:**
NIECAD (SSRAID, SSREID, IAWERR)

☞ **Parameters:**
Input:

SSRAID   identifier of the set of administrative data of the assembly (owner) from which the element (member) is to be excluded, disconnected

SSREID   identifier of the set of administrative data of the element (member) to be excluded, disconnected from the given assembly (owner)

Output:
IAWERR   error status

☞ **Function:**
Disconnection of the set of administrative data of the element (member) from the set of administrative data of the assembly (owner). By this function the element will not be deleted. If the member is a global set of administrative data it exists furthermore as an independent model in the database. If the member is a local one it must be deleted or included into another assembly. The user needs access right WRITE or DELETE for the set of administrative data which is the owner of the excluding element or for any owner of it.

☞ **Error Messages:**

Kind	No.	Message
1	107	NI: the element to be included does not exist
1	109	NI: the assembly does not possess the element
3	110	NI: the excluded element was the last one of the assembly
1	111	NI: set of administrative data with the given identifier does not exist
1	213	NI: identifier of assembly (owner) is empty

| 1 | 214 | NI: identifier of data set to be included is empty |
| 1 | 216 | NI: user not allowed to include/update data set |

☞ Example:

Starting from the assembly CAR identified by '871224111111' the subassembly DOOR identified by '871223000000' is to be substituted by a new version, the subassembly DOOR identified by'880101222222'. First the old version must be disconnected and then the new version must be included:

CALL NIECAD ('871224111111', '871223000000', ISTAT);

CALL NIICAD ('871224111111', '880101222222', ISTAT);

Fig. 4.2   Example for including a new version

d)   **Select set of administrative data**

☞ **Procedure:**

NISLAD (SSRADA, ISWCUR, IAWERR)

☞ **Parameters:**

Input:

SSRADA   string with the structure of administrative data containing the select criterion; input fields are all fields of SSRADA.

Output:

ISWCUR   number of the cursor referencing the buffer which contains the selected set(s) of administrative data

IAWERR   error status

☞ **Function:**

Selection of all sets of administrative data which fulfill the given select criterion and which the user is allowed to read. In this step all sets of administrative data will be examined, independent of whether they are assemblies or single, stand-alone sets. To list all elements of a given assembly use the function 'LIST ELEMENTS OF AN ASSEMBLY' (look at NILIAD). An empty list of select criteria selects all sets the user is allowed to read.

The user is allowed to read a set of administrative data, if he possesses an access right (read, write or delete) for this set or any owner of it (overlying assemblies). Via the cursor, which is an output of this routine, the user is able to fetch the selected sets from the buffer one after the other by the procedure READ BUFFER (look at NIREAD).

☞ **Remark:**

The given select criteria will be connected by a logical AND, e.g. select all sets of administrative data where CAD/CAM system = 'STRIM' AND creator = 'Miller'.

It is also possible to use wildcards (%); e.g. the select criterion creator = 'Mi%' selects all sets of administrative data where the name of the creator begins with 'Mi'...

e)   **List (select) elements of an assembly at the next lower level**

☞   **Procedure:**
NILIAD (SSRIDF, ISWCUR, IAWERR)

☞   **Parameters:**
Input:

SSRIDF   identifier of the set of administrative data belonging to the assembly, elements of which are to be selected at the next lower level of the assembly structure

Output:

ISWCUR   number of the cursor referencing the buffer which contains the selected set(s) of administrative data

IAWERR   error status

☞   **Function:**
Selection of all sets of administrative data that belong to the given assembly (identified by SSRIDF) and are placed on the next lower level. By this procedure the user is able to navigate level by level through a complex- structured assembly. The user is allowed to list the elements of an assembly if he is allowed to read this assembly or any owner of it.

☞   **Remark:**
The identifier SSRIDF is result of the select-routine NISLAD.

☞   **Error Messages:**

Kind	No.	Message
2	104	NI: the input string is empty/identifier does not exist
2	105	NI: no further elements because the identifier is not one of an assembly
1	111	NI: set of administrative data with the given identifier does not exist
1	233	NI: user is not allowed to select the set of administrative data with the given identifier

f)   **Update set of administrative data**

☞   **Procedure:**
NIUPAD (SSRADA, SSRCON, IAWERR)

☞   **Parameters:**
Input:

SSRADA   string with the structure of administrative data containing the new values in that fields, that shall be updated and the identifier of the set of administrative data, that shall be updated in the database

SADMRS   valid release status of model

SADMRD   release status date of the model

SADMIC   in change

SADMVD   valid until

SADREM   remarks

SADTIT   title

SADLGA   local/global administrative data set and the company specific data

SADEST   engineering status

SADPOD   place of deposit

SADDNR   drawing number

SADDCI   drawing change index

SADSNR   sheet number

SSRCON   string that contains for each field in the set of administrative data the
          information, whether the content of the field has been changed or not (a
          kind of trigger)

Output:
IAWERR   error status

☞ **Function:**
The set of administrative data with the given identifier will be updated with the values
contained in SSRADA. Before the update there will be a consistency check concerning
the new values. The user needs access right WRITE or DELETE for this set of
administrative data or for any owner of it.

☞ **Error Messages:**

Kind	No.	Messager
1	111	NI: set of administrative data with the given identifier does not exist
1	216	NI: user not allowed to include/update data set
2	219	NI: there is no field to be changed
2	220	NI: date of creation of administrative data set must not be changed
2	221	NI: time of creation of administrative data set must not be changed
2	222	NI: user-id of the creator of the administrative data set must not be changed
2	223	NI: receiving date of administrative data set must not be changed
2	224	NI: receiving time of administrative data set must not be changed
2	225	NI: date the model was changed must not be changed
2	226	NI: time the model was changed must not be changed
2	227	NI: user who changed model must not be changed
2	228	NI: date of changing administrative data set must not be changed
2	229	NI: time of changing administrative data set must not be changed
2	230	NI: user who changed administrative data set must not be changed
2	231	CAD/CAM system name must not be changed
1	232	NI: number of element in an assembly must not be changed

☞ **Example:**
In the set of administrative data with the identifier '871224111111' the CAD/CAM
system release is to be changed from 'V1R1' to 'V2R1'

SSRADA at first must be blank
    SSRADA = ' '

the identifier must be assigned
    SADIDF = '871224111111'

the new value must be assigned
    SADCSR = 'V2R1'

assigning the trigger
    only the field CAD/CAM system release is to be changed;
    this is the third field in the string SSRADA, therefore the trigger must be in the
    third character (position) of SSRCON
    SSRCON = '0'
    SSRCON (3:3) = '1';

update the set of administrative data
    CALL NIUPAD (SSRADA, SSRCON, ISTAT)

g.)   **Delete set of administrative data**

☞   **Procedure:**
NIDLAD (SSRIDF, IAWERR)

☞   **Parameters:**
Input:
SSRIDF   identifier of the set of administrative data to be deleted

Output:
IAWERR   error status

☞   **Function:**
Deletes the set of administrative data identified by SSRIDF. Also all access rights and
the model belonging to this set of administrative data will be deleted. In case of an
assembly, the whole structure with all sets of local administrative data and the
corresponding access rights and models will be deleted. Global sets of administrative
data will only be excluded but not deleted. The user needs access right DELETE for
this set of administrative data or for any owner of it (Fig. 4.3).

☞   **Error Messages:**

Kind	No.	Message
2	104	NI: the input string is empty/identifier does not exist
1	111	NI: set of administrative data with the given identifier does not exist
1	211	NI: user is not allowed to delete administrative data set
2	212	NI: warning no closed models deleted
1	234	NI: admministrative data set is still used in other assemblies

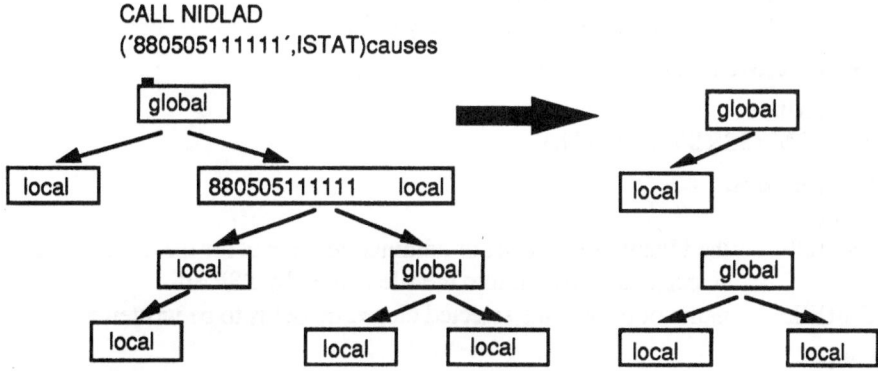

Fig. 4.3   Example for the deletion of administrative data

4.3 Functionality of Geometric Routines

4.3.1 Functions for the Entity CLOSED MODEL

a)   **Insert closed model**

☞   **Procedure:**
NIINCM (SSRIDF, SSRFIL, IAWERR)

☞   **Parameters:**
Input:
SSRIDF   the identifier of the set of administrative data to which the closed model to be inserted belongs
SSRFIL   completely defined filename (depending on the operating system) of the file which contains the closed model in any format (the record length can be either fixed or variable)

Output:
IAWERR   error status

☞   **Function:**
Inserting the closed model into the database by splitting it into single database table rows.

☞   **Remark:**
The application is responsible for the splitting of the closed model into single blocks and several inserts (this means several rows in the data base table).

☞   **Error Messages:**

Kind	No.	Message
2	104	NI: the input string is empty/identifier does not exist
1	111	NI: set of administrative data with the given identifier does not exist
1	118	NI: ascending number is empty
1	119	NI: number of records is empty

1   120   NI: there are no models to be inserted
1   121   NI: closed model already exists

b) Select closed model

☞ **Procedure:**
NISLCM (SSRIDF, SSRFIL)

☞ **Parameters:**
Input:
SSRIDF   the identifier of the set of administrative data for the closed model to be
              selected and written into the file named by SSRFIL
SSRFIL   name of the file the selected closed model is to be written into

Output:
IAWERR   error status

☞ **Function:**
Selection of the closed model, described by the given set of administrative data,
restoring the original format and writing it into the file with the given file name. Only a
user who is allowed to write the corresponding set of administrative data is allowed to
select the closed model, because then he is able to change the geometry in the closed
model (in his CAD/CAM-system).

☞ **Error Messages:**

Kind	No.	Message
2	104	NI: the input string is empty/identifier does not exist
1	111	NI: set of administrative data with the given identifier does not exist

c) Delete closed model

☞ **Procedure:**
NIDLCM (SSRIDF, IAWERR)

☞ **Parameters:**
Input:
SSRIDF   identifier of the set of administrative data of which the closed model is to
              be deleted

Output:
IAWERR   error status

☞ **Function:**
Delete the whole closed model described by the set of administrative data which is
identified by the given identifier. Only a user who is allowed to delete the
corresponding set of administrative data is allowed to delete the closed model.

☞ **Error Messages:**

Kind	No.	Message
1	111	NI: set of administrative data with the given identifier does not exist

4.3.2   Functions for the Entity B_REP

a)   **Insert B_Rep**

☞   **Procedure:**
NIINBR (CSRNAM,ISWNAM,IAWERR)

☞   **Parameters:**
Input:
CSRNAM   user_defined name of the B_Rep

Output:
ISWNAM   contains the internal name of the B_Rep generated

IAWERR   error status of the routine

☞   **Function:**
Insert an empty entity B_Rep into the database in the scope of the enclosing entity.

☞   **Error Messages:**

Kind	No.	Message
1	1	NI: there is no opened scope
1	2	DB: table full, can't insert
1	3	DB: not enough space in the database, can't insert
1	16	NI: B_Rep name is empty
1	17	NI: B_Rep name is to long
1	18	NI: B_Rep name already exists

b)   **Select B_Rep**

☞   **Procedure:**
NISLBR (CSRNAM,ISWNAM,IAWERR)

☞   **Parameters:**
Input:
CSRNAM   user defined name of the B_Rep

Output:
ISWNAM   contains the internal name of the B_Rep-entity
IAWERR   error status of the routine

☞   **Function:**
Selects B_Rep from the database from the scope of the enclosing entity.

☞   **Error Messages:**

Kind	No.	Message
1	1	NI: there is no opened scope
3	9	NI: name found in an outer scope
1	16	NI: B_Rep name is empty
1	17	NI: B_Rep name is to long
1	19	NI: B_Rep name not found

c)   **Update b_rep**

☞   **Procedure:**
NIUPBR (CSRNAM,CSRNEW,ISWNAM,IAWERR)

☞ **Parameters:**

Input:
CSRNAM   old user defined name of the B_Rep
CSRNEW   new user defined name of the B_Rep

Output:
ISWNAM   contains the internal name of the B_Rep-entity
IAWERR   Error status of the routine

☞ **Function:**
Updates B_Rep in the database in the scope of the enclosing entity.

☞ **Error Messages:**

Kind	No.	Message
1	1	NI: there is no opened scope
1	16	NI: B_Rep name is empty
1	17	NI: B_Rep name is to long
1	18	NI: B_Rep name already exists
1	19	NI: B_Rep name not found

d)   **Delete b_rep**

☞ **Procedure:**
NIDLBR (CSRNAM,IAWERR)

☞ **Parameters:**

Input:
CSRNAM   user name of the B_Rep

Output:
IAWERR   error status of the routine

☞ **Function:**
Deletes B_Rep in the database in the scope of the enclosing entity.

☞ **Error Messages:**

Kind	No.	Message
1	1	NI: there is no opened scope
1	16	NI: B_Rep name is empty
1	17	NI: B_Rep name is to long
1	19	NI: B_Rep name not found
1	20	NI: Unable to delete an opened B_Rep

e)   **List B_Rep**

☞ **Procedure:**
NILIBR (ISRNAM,IAWCON,IAWERR)

☞ **Parameters:**

Input:
ISRNAM   name of the B_Rep

Output:
IAWCON   array containing statistical information about the content of the B_Rep
IAWERR   error status of the routine

☞ **Function:**
Lists statistical information about the content of the B-Rep.

☞ **Error Messages:**

KIND	NR	MESSAGE
1	1	NI: there is no opened scope
1	8	NI: name not found
1	7	NI: invalid name for an entity, only >0 allowed
1	23	NI: No B_Rep opened to list statistics
2	24	NI: B_Rep is empty

f) **Open B_Rep**

☞ **Procedure:**
NIOPBR (ISRNAM,ISWNUM,ISRSNM,IAWSHL,IAWERR)

☞ **Parameters:**
Input:
ISRNAM: name of the B_Rep

ISRSNM: maximum number of values that may be returned by IAWSHL

Output:
ISWNUM: contains the number of shell(s) of the B_Rep
IAWSHL: array containing the names of shell(s) of the B_Rep
IAWERR: error status of the routine

☞ **Function:**
opens B_Rep, makes the scope of the B-Rep the actual scope.

☞ **Error Messages:**

Kind	No.	Message
1	1	NI: there is no opened scope
1	14	NI: field overflow
1	16	NI: B_Rep name is empty
1	17	NI: B_Rep name is to long
1	19	NI: B_Rep name not found
1	21	NI: another B_Rep is still open, please close first

g) **Close B_Rep**

☞ **Procedure:**
NICLBR (ISRNAM,IAWERR)

☞ **Parameters:**
Input:
ISRNAM: name of the B_Rep

Output:
IAWERR: error status of the routine

☞ **Function:**
Closes the B_Rep in the database in the scope of the enclosing entity.

☞ **Error Messages:**

Kind	No.	Message
1	1	NI: there is no opened scope

1	8	NI: name not found
1	22	NI: no B_Rep opened to close
2	27	NI: remaining unused Attributes have been deleted from the scope of the closed entity

4.3.3 Functions for the Entity Shell

a) **Insert Shell**

☞ **Procedure:**
NIINSH (ISRNAM,ISRFNM,IARFAC,IAWERR)

☞ **Parameters:**
Input:

ISRNAM	name of the shell entity
ISRFNM	number of faces of the shell to be inserted
IARFAC	array containing the referenced faces of the shell

Output:

IAWERR	error status of the routine

☞ **Function:**
Insert shell into the database in the scope of the enclosing entity.

☞ **Error Messages:**

Kind	No.	Message
1	1	NI: there is no opened scope
1	2	DB: table full, can't insert
1	3	DB: not enough space in the database, can't insert
1	6	NI: entity with this Name already exists in the scope
1	7	NI: invalid name for an entity, only >0 allowed
1	8	NI: name not found
1	10	NI: name out of range
1	13	NI: nothing to insert, number of elements <1

b) **Select Shell**

☞ **Procedure:**
NISLSH (ISRNAM,ISRFNM,ISWBRU,ISWFNM,IAWFAC,IAWERR)

☞ **Parameters:**
Input:

ISRNAM	name of the shell
ISRFNM	maximum number of values that may be returned by IAWFAC

Output:

ISWBRU	internal name of the B_Rep
ISWFNM	number of selected faces of the shell
IAWFAC	array containing the referenced faces of the shell
IAWERR	error status of the routine

☞ **Function:**
Selects shell from the database from the scope of the enclosing entity.

☞ **Error Messages:**

Kind	No.	Message
1	1	NI: there is no opened scope
1	10	NI: name out of range
1	8	NI: name not found
3	9	NI: name found in an outer scope
1	14	NI: field overflow

c) **Update Shell**

☞ **Procedure:**
NIUPSH (ISRNAM,ISRFNM,IARFAC,IAWERR)

☞ **Parameters:**
Input:
ISRNAM   name of the shell
ISRFNM   number of faces of the shell to be updated
IARFAC   array containing the referenced faces of the shell

Output:
IAWERR   error status of the routine

☞ **Function:**
Updates shell in the database in the scope of the enclosing entity.

☞ **Error Messages:**

Kind	No.	Message
1	1	NI: there is no opened scope
1	10	NI: name out of range
1	8	NI: name not found
3	9	NI: name found in an outer scope
1	15	NI: nothing to update, number of elements <1

d) **Delete Shell**

☞ **Procedure:**
NIDLSH (ISRNAM,IAWERR)

☞ **Parameters:**
Input:
ISRNAM   name of the shell

Output:
IAWERR   error status of the routine

☞ **Function:**
Deletes shell in the database in the scope of the enclosing entity.

☞ **Error Messages:**

Kind	No.	Message
1	1	NI: there is no opened scope
1	10	NI: name out of range
1	8	NI: name not found
3	9	NI: name found in an outer scope

4.3.4 Functions for the Entity Face

a) **Insert Face**

☞ **Procedure:**
NIINFA (ISRNAM,ISRPSU,LSRFOR,ISRLNM,IARLOP,IAWERR)

☞ **Parameters:**
Input:

ISRNAM	name of the face
ISRPSU	contains the name of the planar-surface defining the face
LSRFOR	logical orientation of the planar-surface
ISRLNM	number of loop(s) of the face to be inserted
IARLOP	array containing the referenced loops of the face

Output:

IAWERR	error status of the routine

☞ **Function:**
Inserts face in the database in the scope of the enclosing entity.

☞ **Error Messages:**

Kind	No.	Message
1	1	NI: there is no opened scope
1	2	DB: table full, can't insert
1	3	DB: not enough space in the database, can't insert
1	6	NI: entity with this Name already exists in the
1	7	NI: invalid name for an entity, only >0 allowed
1	8	NI: name not found
1	10	NI: name out of range
1	13	NI: nothing to insert, number of elements <1

b) **Select Face**

☞ **Procedure:**
NISLFA (ISRNAM,ISWPSU,LSWFOR,ISRLNM,ISWLNM,IAWLOP,IAWERR)

☞ **Parameters:**
Input:

ISRNAM:	name of the face
ISRLNM:	maximum number of values that may be returned by IAWLOP

Output:

ISWPSU	contains the name of the planar surface defining the face
LSWFOR	logical orientation of the planar surface
ISWLNM	number of the selected loops of the face
IAWLOP	array containing the referenced loops of the face
IAWERR	error status of the routine

☞ **Function:**
Selects face from the database from the scope of the enclosing entity.

☞   **Error Messages:**

Kind	No.	Message
1	1	NI: there is no opened scope
1	10	NI: name out of range
1	8	NI: name not found
3	9	NI: name found in an outer scope
1	14	NI: field overflow

c)   **Update face**

☞   **Procedure:**
NIUPFA (ISRNAM,ISRPSU,LSRFOR,ISRLNM,IARLOP,IAWERR)

☞   **Parameters:**
Input:

ISRNAM	name of the face
ISRPSU	contains the name of the planar-surface defining the face
LSRFOR	logical orientation of the planar-surface
ISRLNM	number of the loops of the face
IARLOP	array containing the referenced loops of the face

Output:

IAWERR	error status of the routine

☞   **Function:**
Updates face in the database in the scope of the enclosing entity.

☞   **Error Messages:**

Kind	No.	Message
1	1	NI: there is no opened scope
1	10	NI: name out of range
1	8	NI: name not found
3	9	NI: name found in an outer scope
1	15	NI: nothing to update, number of elements <1

☞   **Procedure:**
NIDLFA (ISRNAM,IAWERR)

d)   **Delete Face**

☞   **Parameters:**
Input:

ISRNAM	name of the face

Output:

IAWERR	error status of the routine

☞   **Function:**
Deletes face in the database in the scope of the enclosing entity.

☞   **Error Messages:**

Kind	No.	Message
1	1	NI: there is no opened scope
1	10	NI: name out of range

1	8	NI: name not found
3	9	NI: name found in an outer scope

4.3.5 Functions for the Entity EDGE_LOOP

a)   Insert Loop

☞ **Procedure:**
NIINLP (ISRNAM,ISRENM,IAREDR,LAREDO,IAWERR)

☞ **Parameters:**
Input:

ISRNAM	name of the loop
ISRENM	number of edges of the loop to be inserted
IAREDR	array containing the referenced edges of the loop
LAREDO	array containing the orientation of the referenced edges of the loop

Output:

IAWERR	error status of the routine

☞ **Function:**
Inserts loop in the database in the scope of the enclosing entity.

☞ **Error Messages:**

Kind	No.	Message
1	1	NI: there is no opened scope
1	2	DB: table full , can't insert
1	3	DB: not enough space in the database, can't insert
1	5	NI: integer-value out of range
1	6	NI: entity with this name already exists in the scope
1	7	NI: invalid name for an entity, only >0 allowed
1	8	NI: name not found
1	10	NI: name out of range
1	13	NI: nothing to insert, number of elements <1

b)   Select loop

☞ **Procedure:**
NISLLP (ISRNAM,ISRENM,ISWENM,IAWEDR,LAWEDO,IAWERR)

☞ **Parameters:**
Input:

ISRNAM	name of the loop
SRENM	maximum number of values that may be returned by IAWEDR and LAWEDO

Output:

ISWENM	number of selected edges of the selected loop
IAWEDR	array containing the referenced edges of the loop
LAWEDO	array containing the orientation of the referenced edges of loop elements
IAWERR	error status of the routine

☞ **Function:**
Selects loop from the database from the scope of the enclosing entity.

☞ **Error Messages:**

Kind	No.	Message
1	1	NI: there is no opened scope
1	10	NI: name out of range
1	8	NI: name not found
3	9	NI: name found in an outer scope
1	14	NI: field overflow

c)   **Update Loop**

☞ **Procedure:**
NIUPLP (ISRNAM,ISRENM,IAREDR,LAREDO,IAWERR)

☞ **Parameters:**
Input:

ISRNAM	name of the loop
ISRENM	number of edges of the loop to be updated
IAREDR	array containing the referenced edges of the loop
LAREDO	array containing the orientation of the referenced edges of the loop

Output:
IAWERR   error status of the routine

☞ **Function:**
Updates loop in the database in the scope of the enclosing entity.

☞ **Error Messages:**

Kind	No.	Message
1	1	NI: there is no opened scope
1	10	NI: name out of range
1	8	NI: name not found
3	9	NI: name found in an outer scope
1	15	NI: nothing to update, number of elements <1

d)   **Delete Loop**

☞ **Procedure:**
NIDLLP (ISRNAM,IAWERR)

☞ **Parameters:**
Input:
ISRNAM   name of the loop

Output:
IAWERR   error status of the routine

☞ **Function:**
Deletes loop in the database in the scope of the enclosing entity.

☞ **Error Messages:**

Kind	No.	Message
1	1	NI: there is no opened scope
1	10	NI: name out of range
1	8	NI: name not found

4.3.6   Functions for the Entity EDGE

a)   Insert Edge

☞   **Procedure:**
NIINED (ISRNAM,ISRLIN,ISRSVE,ISREVE,IAWERR)

☞   **Parameters:**
Input:

ISRNAM	name of the edge
ISRLIN	contains the name of the line defining the edge
ISRSVE	contains the name of the start-vertex defining the edge
ISREVE	contains the name of the end-vertex defining the edge

Output:

IAWERR	error status of the routine

☞   **Function:**
Inserts edge in the database in the scope of the enclosing entity.

☞   **Error Messages:**

Kind	No.	Message
1	1	NI: there is no opened scope
1	2	DB: table full, can't insert
1	3	DB: not enough space in the database, can't insert
1	5	NI: integer value out of range
1	6	NI: entity with this name already exists in the table
1	7	NI: invalid name for an entity, only >0 allowed
1	8	NI: name not found
1	10	NI: name out of range

b)   Select Edge

☞   **Procedure:**
NISLED (ISRNAM,ISWLIN,ISWSVE,ISREVE,IAWERR)

☞   **Parameters:**
Input:

ISRNAM	name of the edge

Output:

ISWLIN	contains the name of the line-entity defining the edge
ÏSWSVE	contains the name of the start-vertex defining the edge
ISWEVE	contains the name of the end-vertex defining the edge
IAWERR	error status of the routine

☞   **Function:**
Selects edge from the database from the scope of the enclosing entity.

☞   **Error Messages:**

Kind	No.	Message
1	1	NI: there is no opened scope
1	10	NI: name out of range

1	8	NI: name not found
3	9	NI: name found in an outer scope

c) Update Edge

☞ **Procedure:**
NIUPED (ISRNAM,ISRLIN,ISRSVE,ISREVE,IAWERR)

☞ **Parameters:**
Input:

ISRNAM	name of the edge
ISRLIN	contains the name of the line defining the edge
ISRSVE	contains the name of the start-vertex defining the edge
ISREVE	contains the name of the end-vertex defining the edge

Output:

IAWERR	error status of the routine

☞ **Function:**
Updates edge in the database in the scope of the enclosing entity.

☞ **Error Messages:**

Kind	No.	Message
1	1	NI: there is no opened scope
1	10	NI: name out of range
1	8	NI: name not found
3	9	NI: name found in an outer scope

d) Delete Edge

☞ **Procedure:**
NIDLED (ISRNAM,IAWERR)

☞ **Parameters:**
Input:

ISRNAM	name of the edge

Output:

IAWERR	error status of the routine

☞ **Function:**
Deletes edge in the database in the scope of the enclosing entity.

☞ **Error Messages:**

Kind	No.	Message
1	1	NI: there is no opened scope
1	10	NI: name out of range
1	8	NI: name not found

4.3.7 Functions for the Entity Vertex

a) Insert Vertex

☞ **Procedure:**
NIINVE   (ISRNAM,ISRXVE,RSRXVE,ISRYVE,
              RSRYVE,ISRZVE,RSRZVE,IAWERR)

☞   **Parameters:**

Input:

ISRNAM   name of the vertex

ISRXVE   contains the name of a real entity defining the X-coordinate of the
vertex(if ISRXVE is not equal to zero)

RSRXVE   contains the value defining the X-coordinate of the vertex (if ISRXVE is
equal to zero)

ISRYVE   contains the name of a real entity defining the X-coordinate of the vertex
(if ISRYVE is not equal to zero)

RSRYVE   contains the value defining the X-coordinate of the vertex (if ISRYVE is
equal to zero)

ISRZVE   contains the name of a real entity defining the Z-coordinate of the vertex
(if ISRZVE is not equal to zero)

RSRZVE   contains the value defining the Z-coordinate of the vertex (if ISRZVE is
equal to zero)

Output:

IAWERR: error status of the routine

☞   **Function:**

Insert vertex in the database in the scope of the enclosing entity.

☞   **Error Messages:**

Kind	No.	Message
1	1	NI: there is no opened scope
1	2	DB: table full, can't insert
1	3	DB: not enough space in the database, can't insert
1	5	NI: integer value out of range
1	6	NI: entity with this Name already exists in the table
1	7	NI: invalid name for an entity, only >0 allowed
1	8	NI: name not found
1	10	NI: name out of range

b)   **Select Vertex**

☞   **Procedure:**

NISLVE   (ISRNAM,ISWXVE,RSWXVE,ISWYVE,
RSWYVE,ISWZVE,RSWZVE,IAWERR)

☞   **Parameters:**

Input:

ISRNAM   name of the vertex

Output:

ISWXVE   contains the name of a real entity defining the X-coordinate of the vertex
(if ISWXVE is not equal to zero)

RSWXVE   contains the evaluated value defining the X-coordinate of the vertex

ISWYVE   contains the name of a real entity defining the X-coordinate of the vertex
(if ISWYVE is not equal to zero)

RSWYVE   contains the evaluated value defining the X-coordinate of the vertex

ISWZVE   contains the name of a real entity defining the Z-coordinate of the vertex
(if ISWZVE is not equal to zero)

RSWZVE   contains the evaluated value defining the Z-coordinate of the vertex
IAWERR   error status of the routine

☞ **Function:**
Selects vertex from the database from the scope of the enclosing entity.

☞ **Remark:**
The coordinates of the vertex are described in the table of vertex (no references to a name of real entity).

☞ **Error Messages:**

Kind	No.	Message
1	1	NI: there is no opened scope
1	10	NI: name out of range
1	8	NI: name not found
3	9	NI: name found in an outer scope

c)   **Update vertex**

☞ **Procedure:**
NIUPVE   (ISRNAM,ISRXVE,RSRXVE,ISRYVE,
          RSRYVE,ISRZVE,RSRZVE,IAWERR)

☞ **Parameters:**
Input:
ISRNAM   name of the vertex
ISRXVE   contains the new name of a real entity defining the X-coordinate of the vertex (if ISRXVE is not equal to zero)
RSRXVE   contains the new value defining the X-coordinate of the vertex (if ISRXVE is equal to zero)
ISRYVE   contains the new name of a real entity defining the X-coordinate of the vertex (if ISRYVE is not equal to zero)
RSRYVE   contains the new value defining the X-coordinate of the vertex (if ISRYVE is equal to zero)
ISRZVE   contains the new name of a real entity defining the Z-coordinate of the vertex (if ISRZVE is not equal to zero)
RSRZVE   contains the new value defining the Z-coordinate of the vertex (if ISRZVE is equal to zero)

Output:
IAWERR   error status of the routine

☞ **Function:**
Updates vertex in the database in the scope of the enclosing entity.

☞ **Error Messages:**

Kind	No.	Message
1	1	NI: there is no opened scope
1	10	NI: name out of range
1	8	NI: name not found
3	9	NI: name found in an outer scope

d) **Delete Vertex**

☞ **Procedure:**
NIDLVE (ISRNAM,IAWERR)

☞ **Parameters:**
Input:
ISRNAM   name of the vertex

Output:
IAWERR   error status of the routine

☞ **Function:**
Deletes vertex in the database in the scope of the enclosing entity.

☞ **Error Messages:**

Kind	No.	Message
1	1	NI: there is no opened scope
1	10	NI: name out of range
1	8	NI: name not found

4.3.8 Functions for the Entity Planar Surface

a) **Insert Planar Surface**

☞ **Procedure:**
NIINPS   (ISRNAM,ISRPNT,RARPNT,ISRDIR,
            RARDIR,ISRREF,RARREF,IAWERR)

☞ **Parameters:**
Input:
ISRNA   name of the planar surface
ISRPNT   contains the name of the point defining the planar surface
RARPNT   array containing the components of the point
ISRDIR   contains the name of the normal-direction defining the planar surface
RARDIR   array containing the components of the direction
ISRREF   contains the name of the reference point defining the planar surface
RARREF   array containing the components of the reference point

Output:
IAWERR   error status of the routine

☞ **Function:**
Inserts planar surface in the database in the scope of the enclosing entity.

☞ **Error Messages:**

Kind	No.	Message
1	1	NI: there is no opened scope
1	2	DB: table full, can't insert
1	3	DB: not enough space in the database, can't insert
1	5	NI: integer value out of range
1	6	NI: entity with this name already exists in the
1	7	NI: invalid name for an entity, only >0 allowed

1	8	NI: name not found
1	10	NI: name out of range

b)   **Select Planar Surface**

☞   **Procedure:**
NISLPS   (ISRNAM,ISWPNT,RAWPNT,ISWDIR,
            RAWDIR,ISRREF,RAWREF,IAWERR)

☞   **Parameters:**
Input:
ISRNAM   name of the planar_surface

Output:
ISWPNT   contains the name of the point defining the planar surface
RAWPNT   array containing the components of the point
ISWDIR   contains the name of the normal-direction defining the planar surface
RAWDIR   array containing the components of the direction
ISWREF   contains the name of the reference point defining the planar surface
RAWREF   array containing the components of the reference point
IAWERR   error status of the routine

☞   **Function:**
Selects planar surface from the database from the scope of the enclosing entity.

☞   **Error Messages:**

Kind	No.	Message
1	1	NI: there is no opened scope
1	10	NI: name out of range
1	8	NI: name not found
3	9	NI: name found in an outer scope

c)   **Update Planar Surface**

☞   **Procedure:**
NIUPPS   (ISRNAM,ISRPNT,RARPNT,ISRDIR,
            RARDIR,ISRREF,RARREF,IAWERR)

☞   **Parameters:**
Input:
ISRNAM   name of the planar surface
ISRPNT   contains the name of the point defining the planar surface
RARPNT   array containing the components of the point
ISRDIR   contains the name of the normal-direction defining the planar surface
RARDIR   array containing the components of the direction
ISRREF   contains the name of the reference point defining the planar surface
RARREF   array containing the components of the reference point

Output:
IAWERR: error status of the routine

☞   **Function:**
Updates planar_surface in the database in the scope of the enclosing entity.

☞ **Error Messages:**

Kind	No.	Message
1	1	NI: there is no opened scope
1	10	NI: name out of range
1	8	NI: name not found
3	9	NI: name found in an outer scope

d) **Delete Planar Surface**

☞ **Procedure:**
NIDLPS (ISRNAM,IAWERR)

☞ **Parameters:**
Input:
ISRNAM   name of the planar surface

Output:
IAWERR   error status of the routine

☞ **Function:**
Deletes planar surface in the database in the scope of the enclosing entity.

☞ **Error Messages:**

Kind	No.	Message
1	1	NI: there is no opened scope
1	10	NI: name out of range
1	8	NI: name not found

4.3.9 Functions for the Entity Line

a) **Insert Line**

☞ **Procedure:**
NIINLI (ISRNAM,ISRPNT,RARPNT,ISRDIR,RARDIR,IAWERR)

☞ **Parameters:**
Input:
ISRNAM   name of the line
ISRPNT   contains the name of the point defining the line
RARPNT   array containing the coordinates of the point
ISRDIR   contains the name of the direction defining the line
RARDIR   array containing the coordinates of the direction

Output:
IAWERR   error status of the routine

☞ **Function:**
Inserts line in the database in the scope of the enclosing entity.

☞ **Error Messages:**

Kind	No.	Message
1	1	NI: there is no opened scope
1	2	DB: table full, can't inserted
1	3	DB: not enough space in the database, can't insert
1	5	NI: integer value out of range

1	6	NI: entity with this name already exits in the table
1	7	NI: invalid name for an entity, only >0 allowed
1	8	NI: name not found
1	10	NI: name out of range

b)   **Select Line**

☞   **Procedure:**
NISLLI (ISRNAM,ISWPNT,RAWPNT,ISWDIR,RAWDIR,IAWERR)

☞   **Parameters:**
Input:
ISRNAM   name of the line

Output:
ISWPNT   contains the name of the point defining the line
RAWPNT   array containing the coordinates of the point
ISWDIR   contains the name of the direction definingthe line
RAWDIR   array containing the coordinates of the direction
IAWERR   error status of the routine

☞   **Function:**
Selects line from the database from the scope of the enclosing entity.

☞   **Error Messages:**

Kind	No.	Message
1	1	NI: there is no opened scope
1	10	NI: name out of range
1	8	NI: name not found
3	9	NI: name found in an outer scope

c)   **Update Line**

☞   **Procedure:**
NIUPLI (ISRNAM,ISRPNT,RSWPNT,ISRDIR,RSWDIR,IAWERR)

☞   **Parameters:**
Input:
ISRNAM   name of the line
ISRPNT   contains the name of the point defining the line
RARPNT   array containing the coordinates of the point
ISRDIR   contains the name of the direction definingthe line
RARDIR   array containing the coordinates of the direction

Output:
IAWERR   error status of the routine

☞   **Function:**
Updates line in the database in the scope of the enclosing entity.

☞   **Error Messages:**

Kind	No.	Message
1	1	NI: there is no opened scope
1	10	NI: name out of range
1	8	NI: name not found
3	9	NI: name found in an outer scope

d)   Delete Line

☞   **Procedure:**
NIDLLI (ISRNAM,IAWERR)

☞   **Parameters:**
Input:
ISRNAM   name of the line

Output:
IAWERR   error status of the routine

☞   **Function:**
Deletes line in the database in the scope of the enclosing entity.

☞   **Error Messages:**

Kind	No.	Message
1	1	NI: there is no opened scope
1	10	NI: name out of range
1	8	NI: name not found

4.3.10 Functions for the Entity Point

a)   Insert Point

☞   **Procedure:**
NIINPT
(ISRNAM,ISRXPT,RSRXPT,ISRYPT,RSRYPT,ISRZPT,RSRZPT,IAWERR)

☞   **Parameters:**
Input:
ISRNAM   name of the point
ISRXPT   contains the name of a real entity defining the X-coordinate of the point
(if ISRXPT is not equal to zero)
RSRXPT   contains the value defining the X-coordinate of the point (if ISRXPT is
equal to zero)
ISRYPT   contains the name of a real entity defining the Y-coordinate of the point
(if ISRYPT is not equal to zero)
RSRYPT   contains the value defining the Y-coordinate of the point (if ISRYPT is
equal to zero)
ISRZPT   contains the name of a real entity defining the Z-coordinate of the point (if
ISRZPT is not equal to zero)
RSRZPT   contains the value defining the Z-coordinate of the point (if ISRZPT
isequal to zero)

Output:
IAWERR   error status of the routine

☞   **Function:**
Inserts point in the database in the scope of the enclosing entity.

☞ **Error Messages:**

Kind	No.	Message
1	1	NI: there is no opened scope
1	2	DB: table full, can't insert
1	3	DB: not enough space in the database, can't insert I
1	4	NI: real value out of range
1	6	NI: entity with this name already exists in table
1	7	NI: invalid name for an entity, only >0 allowed

b)   **Select Point**

☞ **Procedure:**
NISLPT
(ISRNAM,ISWXPT,RSWXPT,ISWYPT,RSWYPT,ISWZPT,RSWZPT,IAWERR)

☞ **Parameters:**
Input:
ISRNAM   name of the point

Output:
ISWXPT   contains the name of a real entity defining the X-coordinate of the point
           (if ISRWPT is not equal to zero)
RSWXPT   contains the evaluated value defining the X-coordinate of the point
ISWYPT   contains the name of a real entity defining the Y-coordinate of the point
           (if ISWYPT is not equal to zero)
RSWYPT   contains the evaluated value defining the Y-coordinate of the point
ISWZPT   contains the name of a real entity defining the Z-coordinate of the point (if
           ISWZPT is not equal to zero)
RSWZPT   contains the evaluated value defining the Z-coordinate of the point
IAWERR   error status of the routine

☞ **Function:**
Selects point from the database from the scope of the enclosing entity.

☞ **Remark:**
The coordinates of the point are described in the table of points (no references to a
name of a real entity).

☞ **Error Messages:**

Kind	No.	Message
1	1	NI: there is no opened scope
1	10	NI: name out of range
1	8	NI: name not found
3	9	NI: name found in an outer scope

c)   **Update Point**

☞ **Procedure:**
NIUPPT   (ISRNAM,ISRXPT,RSRXPT,ISRYPT,
           RSRYPT,ISRZPT,RSRZPT,IAWERR)

☞ **Parameters:**
Input:
ISRNAM   name of the point

ISRXPT contains the new name of a real entity defining the X-coordinate of the point (if ISRXPT is not equal to zero)

RSRXPT contains the new value defining the X-coordinate of the point (if ISRXPT is equal to zero)

ISRYPT contains the new name of a real entity defining the Y-coordinate of the point (if ISRYPT is not equal to zero)

RSRYPT contains the new value defining the Y-coordinate of the point (if ISRYPT is equal to zero)

ISRZPT contains the new name of a real entity defining the Z-coordinate of the point (if ISRZPT is not equal to zero)

RSRZPT contains the new value defining the Z-coordinate of the point (if ISRZPT is equal to zero)

Output:
IAWERR error status of the routine

☞ **Function:**

Updates point in the database in the scope of the enclosing entity.

☞ **Error Messages:**

Kind	No.	Message
1	1	NI: there is no opened scope
1	4	NI: real value out of range
1	10	NI: name out of range
1	8	NI: name not found

d) Delete Point

☞ **Procedure:**

NIDLPT (ISRNAM,IAWERR)

☞ **Parameters:**

Input:
ISRNAM name of the point

Output:
IAWERR error status of the routine

☞ **Function:**

Deletes point in the database in the scope of the enclosing entity.

☞ **Error Messages:**

Kind	Nr	Message
1	1	NI: there is no opened scope
1	10	NI: name out of range
1	8	NI: name not found

e) List All Points

☞ **Procedure:**

NILIPT (ISRASI,IAWNAM,RAWPTX,RAWPTY,RAWPTZ,ISWANZ,IAWERR)

☞ **Parameters:**

Input:
ISRASI contains the parameter for dimensioning the arrays IAWNAM, RAWPTX, RAWPTY and RAWPTZ

Output:
ISWANZ   number of selected points
IAWNAM   array containing the names of the selected points
RAWPTX   array containing the X-components of selected points
RAWPTY   array containing the Y-components of selected points
RAWPTZ   array containing the Z-components of selected points
IAWERR   error status of the routine

☞   **Function:**
Selects points in the database from the scope of the enclosing entity and returns
number, names and x,y,z-coordinates.

☞   **Error Messages:**

Kind	No.	Message
1	1	NI: There is no opened scope
1	11	NI: Number of selected entries greater than parameter for dimensioning arrays
2	12	NI: Nothing to list

4.3.11 Functions for the Entity Direction

a)   **Insert Direction**

☞   **Procedure:**
NIINDI   (ISRNAM,ISRXDI,RSRXDI,ISRYDI,
                  RSRYDI,ISRZDI,RSRZDI,IAWERR)

☞   **Parameters:**
Input:
ISRNAM   name of the direction
ISRXDI   contains the name of a real entity defining the X-coordinate of the point
                (if ISRXDI is not equal to zero)
RSRXDI   contains the value defining the X-coordinate of the point (if ISRXDI is
                equal to zero)
ISRYDI   contains the name of a real entity defining the Y-coordinate of the point
                (if ISRYDI is not equal to zero)
RSRYDI   contains the value defining the Y-coordinate of the point (if ISRYDI is
                equal to zero)
ISRZDI   contains the name of a real entity defining the Z-coordinate of the point
                (if ISRZDI is not equal to zero)
RSRZDI   contains the value defining the Z-coordinate of the point (if ISRZDI is
                equal to zero)

Output:
IAWERR   error status of the routine

☞   **Function:**
Inserts direction in the database in the scope of the enclosing entity.

☞ **Error Messages:**

Kind	No.	Message
1	1	NI: there is no opened scope
1	2	DB: table full, can't insert
1	3	DB: not enough space in the database, can't insert
1	4	NI: real value out of range
1	6	NI: entity with this name already exists in the table
1	7	NI: invalid name for an entity, only >0 allowed

b) **Select Direction**

☞ **Procedure:**

NISLDI   (ISRNAM,ISWXDI,RSWXDI,ISWYDI,
                 RSWYDI,ISWZDI,RSWZDI,IAWERR)

☞ **Parameters:**

Input:

ISRNAM   name of the direction

Output:

ISWXDI   contains the name of a real entity defining the X-coordinate of the point
               (if ISWXDI is not equal to zero)

RSWXDI   contains the evaluated value defining the X-coordinate of the point

ISWYDI   contains the name of a real entity defining the Y-coordinate of the point
               (if ISWYDI is not equal to zero)

RSWYDI   contains the evaluated value defining the Y-coordinate of the point

ISWZDI   contains the name of a real entity defining the Z-coordinate of the point (if
               ISWZDI is not equal to zero)

RSWZDI   contains the evaluated value defining the Z-coordinate of the point

IAWERR   error status of the routine

☞ **Function:**

Selects direction from the database from the scope of the enclosing entity.

☞ **Remark:**

The coordinates of the direction are described in the table of directions (no reference to
a name of a real entity).

☞ **Error Messages:**

Kind	No.	Message
1	1	NI: there is no opened scope
1	10	NI: name out of range
1	8	NI: name not found
3	9	NI: name found in an outer scope

c) **Update Direction**

☞ **Procedure:**

NIUPDI   (ISRNAM,ISRXDI,RSRXDI,ISRYDI,
                 RSRYDI,ISRZDI,RSRZDI,IAWERR)

☞ **Parameters:**

Input:

ISRNAM   name of the direction

ISRXDI	contains the new name of a real entity defining the X-coordinate of the point (if ISRXDI is not equal to zero)
RSRXDI	contains the new value defining the X-coordinate of the point (if ISRXDI is equal to zero)
ISRYDI	contains the new name of a real entity defining the Y-coordinate of the point (if ISRYDI is not equal to zero)
RSRYDI	contains the new value defining the Y-coordinate of the point (if ISRYDI is equal to zero)
ISRZDI	contains the new name of a real entity defining the Z-coordinate of the point (if ISRZDI is not equal to zero)
RSRZDI	contains the new value defining the Z-coordinate of the point (if ISRZDI is equal to zero)

Output:
IAWERR   error status of the routine

☞ **Function:**
Updates direction in the database in the scope of the enclosing entity.

☞ **Error Messages:**

Kind	No.	Message
1	1	NI: there is no opened scope
1	4	NI: real value out of range
1	10	NI: name out of range
1	8	NI: name not found

d)   **Delete Direction**

☞ **Procedure:**
NIDLDI (ISRNAM,IAWERR)

☞ **Parameters:**
Input:
ISRNAM   name of the direction

Output:
IAWERR   Error status of the routine

☞ **Function:**
Deletes direction in the database in the scope of the enclosing entity.

☞ **Error Messages:**

Kind	No.	Message
1	1	NI: there is no opened scope
1	10	NI: name out of range
1	8	NI: name not found

e)   **List all Directions**

☞ **Procedure:**
NILIDI (ISRASI,IAWNAM,RAWDIX,RAWDIY,RAWDIZ,ISWANZ,IAWERR)

☞ **Parameters:**
Input:
ISRASI   contains the parameter for dimensioning the arrays IAWNAM, RAWDIX, RAWDIY and RAWDIZ

Output:

ISWANZ   number of selected directions

IAWNAM   array containing the names of the selected directions

RAWDIX   array containing the X-components of selected directions

RAWDIY   array containing the Y-components of selected directions

RAWDIZ   array containing the Z-components of selected directions

IAWERR   error status of the routine

☞   **Function:**

Selects directions in the database in the scope of the enclosing entity and returns number, name and x,y,z-coordinates.

☞   **Error Messages:**

Kind	No.	Message
1	1	NI: there is no opened scope
1	11	NI: number of selected entries greater than parameter for dimensioning arrays
2	12	NI: nothing to list

4.4 Functionality of Access Routines

4.4.1 Functions for the Entity ACCESS RIGHT

a)   **Insert access right**

☞   **Procedure:**

NIINAC (SSRACC, IAWERR)

☞   **Parameters:**

Input:

SSRACC   string with the structure of access rights containing all values to be inserted; these must be :

- identifier of the set of administrative data to which the access rights belongs

- identifier of the group/user to which the access right is assigned

- the kind of access

(all other fields of SSRACC will be generated by this procedure)

Output:

IAWERR   error status

☞   **Function:**

Date, time and userid of the insertion will be generated. Then the access right will be inserted into the database. Only a user, who possesses an access right DELETE concerning a set of administrative data, is allowed to insert an access right DELETE for this set of administrative data. To insert an access right READ or WRITE, the user himself needs access right WRITE or DELETE for the corresponding set of administrative data.

☞   **Error Messages:**

Kind	No.	Message
2	104	NI: the input string is empty/identifier does not exist
1	113	NI: the group/user does not exist in the system
1	114	NI: the kind of access must be: R=read, W=write, D=delete
1	115	NI: access right already exists
1	170	OS: error during call for actual userid
1	202	NI: identifier of administrative data in access right is empty
1	203	NI: user/group is empty
1	235	NI: user is not allowed to insert access right

☞   Example:

Mr. Miller (with the userid CAD*I012) is to be allowed to read the set of administrative data stored in SSWADA:

SSRACC at first must be blank
    SSRACC = ' '

assigning the values to the input string,

SADIDF contains the identification of the

set of administrative data (received, e.g., by select)
    SACIDF = SADIDF
    SACIUG = 'CAD*I012'
    SACRWD = 'R'

    CALL NIINAC (SSRACC,ISTAT)

b)   **Select access right**

☞   **Procedure:**
NISLAC (SSRIDF, ISWCUR, IAWERR)

☞   **Parameters:**
Input:
SSRIDF   identifier of the set of administrative data of which the access rights are to be selected

Output:
ISWCUR   number of the cursor referencing the buffer which contains the selected access rights

IAWERR   error status

☞   **Function:**
Selection of all access rights belonging to the given set of administrative data.

☞   **Remark:**
The identifier SSRIDF is result of the select routine NISLAD; via the cursor, which is an output of this routine, the user is able to fetch the selected sets from the buffer one after the other by the procedure READ (see NIREAD).

Only a user who is allowed to select the set of administrative data is allowed to select the corresponding access rights.

☞ **Error Messages:**

Kind	No.	Message
1	104	NI: the input string is empty/identifier does not exist
1	111	NI: set of the administrative data with the given identifier does not exist

☞ **Example:**

Select all access rights belonging to the set of administrative data identified by
'871214111111'

> SSRIDF = '871214111111';
>
> CALL NISLAC (SSRIDF, CURS1, ISTAT)

c) **Update access right**

☞ **Procedure:**

NIUPAC (SSRACC, IAWERR)

☞ **Parameters:**

Input:

SSRACC   string with the structure of access rights containing the identifier of the
set of administrative data the access right belongs to, the userid for which
the access right is valid and the new kind of access (read, write, delete)

Output:

IAWERR   error status

☞ **Function:**

Changing the kind of access (read, write, delete) of an existing access right.

Only a user who possesses the access right DELETE concerning a set of administrative
data, is allowed to change READ or WRITE into DELETE and vice verse to change
DELETE into READ or WRITE of this set of administrative data. Any other update is
allowed by a user who possesses access right WRITE for the corresponding set of
administrative data.

☞ **Error Messages:**

Kind	No.	Message
1	114	NI: the kind of access must be: R=read, W=write, D=delete
1	115	NI: access right already exists
1	116	NI: the given access right does not exist
2	205	NI: the identifier of access rights (= ID of data ID of user/group) or part of it is empty
1	237	NI: there is no access right DELETE for the identifier in the system
1	238	NI: user is not allowed to update access right
1	239	NI: no update because the access right with the given identifier is the last one with DELETE for this administrative data set

☞ **Example:**

Mr. Miller (with the userid CAD*I012), who is allowed to read the set of
administrative data identified by '871214111111' is now required to be able to write it:

SSRACC at first must be blank
>   SSRACC = ' '

assigning the identifier of the access right to be changed
>   SACIDF = '871214111111'
>   SACIUG = 'CAD*I012'

assigning the new kind of access
>   SACRWD = 'w'

>   call NIUPAC (SSRACC, ISTAT)

d.) Delete acess right

☞ **Procedure:**
NIDLAC (SSRACC, IAWERR)

☞ **Parameters:**
Input
SSRACC   string with the structure of access rights containing the access right to be
          deleted, at least the identifier of the access right and
          - the identifier of the set of administrative data to which the access right
            belongs
          - the identifier of the group/user to which the access right is assigned

Output:
IAWERR   error status

☞ **Function:**
Delete the given access right.

Only a user who possesses the access right DELETE concerning a set of administrative
data, is allowed to delete an access right DELETE of this set of administrative data. To
delete any other access right of a set of administrative data the user needs the access
right DELETE or WRITE for this set of administrative data.

☞ **Error Messages:**

Kind	No.	Message
1	116	NI: the given access right does not exist
1	117	NI: no deletion because the access right is the last one of the set of administrative data
2	205	NI: the identifier of access rights (= ID of data ID of user/group) or part of it is empty
1	236	NI: user is not allowed to delete access right
1	237	NI: there is no access right DELETE for the identifier in the system

☞ **Example:**
Mr. Miller (with the userid CAD*I012) may no longer write the set of administrative
data identified by 871214111111:

SSRACC at first must be blank
>   SSRACC = ' '

defining the access right that shall be deleted
>   SACIDF = '871214111111'
>   SACIUG = 'CAD*I012'

delete this access right (there can only be one)

     call NIUPAC (SSRACC, ISTAT)

4.4.2 Functions for the Entity Authority Group

List all users /groups

☞ **Procedure:**
NILIAC (SSRGID, ISWCUR, IAWERR)

☞ **Parameters:**
Input:
SSRGID   groupid of the authority group the next level of users or groups shall be displayed

Output:
ISWCUR   number of the cursor referencing the buffer which contains the selected users/groups
IAWERR   error status

☞ **Function:**
Selection of all users/groups of the next lower level starting with the group given by SSRGID in the hierarchical structure of authority.

☞ **Remark:**
Repeating calls of this procedure enables the pass through of the hierarchy (top down).

☞ **Error Messages:**

Kind	No.	Message
1	113	NI: the group/user does not exist in the system
1	203	NI: user/group is empty

4.4.3 System Administrator Functions

a)   **Insert User/Group**

☞ **Procedure:**
NIINUS (SSRUSG, IAWERR)

☞ **Parameters:**
Input:
SSRUSG   string with the structure of user/group definition containing all values to be inserted; these must be
       - identifier of the user or group
       - the type: 'U', if it is a user
                      'G', if it is a group
       - name of the user (or blank, if it is a group)

Output:
IAWERR   error status

☞ **Function:**
The user defined in the input variable SSRUSG will be inserted into the database, in

the table NI_USER_GROUP_DEF. Only users/groups defined in this table are able to get an access right.

☞ **Error Messages:**

Kind	No.	Message
1	250	NI: type of user/group must be 'U' or 'G'
1	251	NI: name of user must be blank
1	252	NI: userid/groupid must not be blank

☞ **Example:**

Mr. Miller (with the userid CAD*IO12) is to be inserted into the database.

SSRUSG at first must be blank

    SSRUSG = ' '

assigning the values to the input string

    SUSIUG = 'CAD*IO12'

    SUSTYP = 'U'

    SUSUNA = 'MILLER'

    CALL NIINUS (SSRUSG, ISTAT)

b) **Select User/Group**

☞ **Procedure:**

NISLUS (SSRUSG, ISWCUR, IAWERR)

☞ **Parameters:**

Input:

SSRUSG   string with the structure of user/group definition containing the select criterion; input fields are all fields of SSRUSG

Output:

ISWCUR   number of the cursor referencing the buffer which contains the selected set(s) of users/groups

IAWERR   error status

☞ **Function:**

Selection of all users/groups which fulfill the given select criteria. An empty list of select criteria selects all users/groups. This function is able to make every kind of selection.

☞ **Example:**

The system administrator wants to have a list of all users which are defined in his system.

SSRUSG at first must be blank

    SSRUSG = ' '

assigning the values to the input string

    SUSTYP = 'U'

    CALL NISLUS (SSRUSG, ISWCUR; ISTAT)

c)   **Update User/Group**

☞   **Procedure:**
NIUPUS (SSRUSG, IAWERR)

☞   **Parameters:**
Input:
SSRUSG   string with the structure of user/group definition containing the identifier
of the user whose name is to be updated

Output:
IAWERR   error status

☞   **Function:**
Changing the name of a user defined in the system. The user defined in the input
variable SSRUSG with the userid SUSIUG will get a new name.

☞   **Error Messages:**

Kind	No.	Message
1	113	NI: the user/group doesn't exist in the system
1	252	NI: userid/groupid may not be blank
1	253	NI: only username can be changed
1	254	NI: the new username may not be blank
1	255	NI: user/group already exists

☞   **Example:**

Mr. Miller (with the userid CAD*IO12) has married, therefore his wife's name must
be updated in the database with the new username "JOHN":

SSRUSG at first must be blank

SSRUSG = ' '

assigning the values to the input string

SUSIUG = 'CAD*IO12'

SUSTYP = 'U'

SUSUNA = 'JOHN'

CALL NIUPUS (SSRUSG, ISTAT)

d)   **Delete User/Group**

☞   **Procedure:**
NIDLUS (SSRIUG, IAWERR)

☞   **Parameters:**
Input:
SSRIUG   the userid or groupid of the user/group that has to be deleted

Output:
IAWERR   error status

☞   **Function:**
The user, defined with the userid, has to be deleted from the database.

☞ **Error Messages:**

Kind	No.	Message
1	113	NI: the user/group doesn't exist in the system
1	252	NI: userid/groupid may not be blank

☞ **Example:**
Mr. Miller (with the userid CAD*IO12) has to be deleted from the database:

Assigning the values to the input string

    SSRIUG = 'CAD*IO12'

    CALL NIDLUS (SSRIUG, ISTAT)

e) **Insert Father-Son**

☞ **Procedure:**
NIINTR (SSRTRE, IAWERR)

☞ **Parameters:**
Input:
SSRTRE   this string contains the father-son connection to be inserted; it must
              include:
              - identifier of the father group
              - identifier of the son user or the son group

Output:
IAWERR   error status

☞ **Function:**
The user/groups defined in the NI_USER_GROUP_DEF table will be inserted into
the NI_GROUP_TREE table in the database. With this function the system
administrator is able to build up the group tree by inserting father-son connections.

☞ **Error Messages:**

Kind	No.	Message
1	256	NI: id of father/son may not be blank
1	258	NI: a father may not be a user
1	259	NI: a previous father may not be a son
1	260	NI: the father-son connection already exists

☞ **Example:**
Mr. Miller (with the userid CAD*IO12) has to be inserted as a son of the group, with
an groupid CAD, in the database.

Assigning the values to the input string

STRFAT = 'CAD'

STRSON = 'CAD*IO12'

CALL NIINTR (SSRTRE, ISTAT)

f) **Delete Father-Son**

Procedure:
NIDLTR (SSRTRE, IAWERR)

Parameters:

Input:

SSRTRE   this string contains the father-son relation to be deleted; it must include:
- identifier of the father group
- identifier of the son user or son group

Output:

IAWERR   error status

Function:

A father/son pair defined with the userid has to be deleted from the database. The connection between the given father and the given son will be cancelled.

Error Messages:

Kind	No.	Message
1	256	NI: id of father/son may not be blank
1	260	NI: the father-son connection already exists

Example:

Mr. Miller (with the userid CAD*IO12) as a son and the group with the groupid CAD as a father has to be deleted from the database.

Assigning the values to the input string

STRFAT = 'CAD'

STRSON = 'CAD*IO12'

CALL NIDLTR (SSRTRE, ISTAT)

5   Database Schema Definition

5.1 General Rules

☞ The conceptual schema in IDEF1x is used in the CAD*I database group to describe the handled entities, properties and attributes,

☞ The formal data structure used in the CAD*I project is described according to the High Level Data Specification Language (HDSL),

☞ This data schema is divided into two main parts which are the Geometric part(B_REP model with planar faces) and the Administrative part covering the main administrative data used in industry according to the CAD*I database group(administrative data, access rights, closed models, etc.).

5.1.1 How Particular Concepts are Represented in the RDBMS Area

The main concepts the implementation of which is described in this paragraph are those used in the subset of the CAD*I-ORACLE data schema implemented by the CAD*I Database group and detailed below :

☞ the attribute type ANY and REF_ANY

☞ types related to referencing mechanism REFERENCE and REF_ONLY,

☞ the list type LIST OF, enumeration type ENUM and class type CLASS

5.1.1.1 The Attribute Type

The types ANY and REF_ANY indicate that data is given in elementary form or as reference to an entity of the corresponding type. In the database context, the elementary form description is performed in the table describing the data.

Example: POINT treated as an attribute (elementary form)

SCOPE-NUMBER-LI	LI-NAME	LI-POINT	LI-X-PTR	LI-Y-PTR	LIZ-PTR
1	NAMLI	NAMPTR	10.	0.5	0.

		LI-DIRECTION	LI-X-DIR	LI-Y-DIR	LI-Z-DIR
		0	1.	0.	0.

On the other side, any referenced entity is represented by a relationship between two tables. The first table describes the data and the second table the referenced entity.

The relationship between such tables is performed via the name of the referenced entity.

Example : POINT treated as a referenced entity in the planar-surface table

First table (NI_PLANAR_SURFACE) :

SCOPE-NUMBER-PS	PS-NAME	PS-POINT	PS-X-PTR	PS-Y-PTR	PS-Z-PTR
1	NAMPS	NAMPNT	0.	0.	0.

		PS-DIRECTION	PS-X-DIR	PS-Y-DIR	PS-Z-DIR
		10	1.	0.	0.

		PS-REF-POINT	PS-X-REF	PS-Y-REF	PS-Z-REF
		5	10.	0.	3.2

Second table (NI_POINT) :

SCOPE-NUMBER-PT	POINT-NAME	X-NAME-PT	X-COMPONENT-PT
1	NAMPNT	0	0.5

		Y-NAME-PT	Y-COMPONENT-PT
		0	0.

		Z-NAME-PT	Z-COMPONENT-PT
		0	10.3

5.1.1.2 Types Related to Referencing Mechanism

We have seen in the previous paragraph that entities may be referenced. Indeed, all relationships between entities are expressed by references. Two types of referencing mechanisms are provided and used mainly in the CAD*I data schema implemented by the CAD*I Database group: REFERENCE and REF_ONLY. These allow for either multiple references to the same entity(1-m), or only single reference to an entity(1-1). For example, the face entity described in the CAD*I data schema references a Planar_surface entity according to a single reference.

Therefore, in the CAD*I-ORACLE database schema implemented the face table contains one and only one reference to a name of a Planar_surface entity and similarly the Planar_surface entity (located by its name) appears in one and only one line of the face table.

5.1.1.3 The List Type, Enumeration Type and Class Type

The list type is a finite and arbitrary collection of elements. elements of a list have to be of the same type. This type is a dynamic type in the sense that the number of elements in the list is variable. In the CAD*I-ORACLE data schema implemented, this type is mainly used for entities described by a list of reference to other entities.

In the database area, the representation of such a type is based on the utilization of two tables in relationship between each other by a unique identifier. The second table contains the "list of references" to other entities describing the entity of the first table.

Example: tables related to the FACE entity

First table (NI_FACE) :

SCOPE_NUMBER_FA	FACE_NAME	PS_NAME	FA_PS_ORIENT	FACE_IDENT
1	NAMFA	NAMPS	1	10

Second table (NI_FACE_HELP) : This table references all the LOOP entities defining the FACE.

FACE_IDENTIFIER	FA_LOOP_ORDER	FA_LOOP_REF
10	1	3
10	2	6
10	3	9
10	4	12

The last two types (enumeration and class types) are used in the CAD*I data schema to describe a finite enumeration of constants or types.

In the CAD*I database schema implemented, such types are used to define constraints on attribute values of one entity. For example, the AUTHORITY entity describes constraints related to accesses allowed to administrative data of any model :

- READ, WRITE and DELETE.

5.2 Definition of the Tables

5.2.1 The Administrative Data Table

This table contains the administrative data. It consists of the general part and all specific parts. Regarding a part geometry, only the columns of the general part and the columns of the geometry-specific part have values, but the columns of the other specific parts are null.

TABLE_NAME :   NI_ADM_DATA

Type	Column	Data Type	Key	Description
M	SADIDF	char*14	P	identification of the set of administrative data
	SADCSN	char* 10		CAD/CAM_system name
	SADCSR	char* 8		CAD/CAM_system_release
	SADDAC	char* 6		data change of administrative data
	SADTAC	char* 6		time change of administrative data

	SADCID	char* 20	userid of user who changed administrative data
	SADADC	char* 6	date of creation
	SADATC	char* 6	time of creation
	SADUIC	char* 20	userid of user who created the model
	SADADR	char* 6	date when the model and its adm. data were received (input) by the database
	SADATR	char* 6	time when the model and its adm. data were received (input) by the database
	SADDMC	char* 6	date when model was changed
	SADTMC	char* 6	timewhen model was changed
	SADMUC	char* 20	userid of user who changed model
	SADHOT	char* 20	host_type
	SADHOS	char* 10	host_operating_system the model comes from
	SADHOR	char* 8	host_operating_system_release the model comes from
	SADTOM	char* 20	type of model (part- or tool-geometry)
	SADVAR	char* 10	variant
	SADVER	char* 3	version
	SADMRS	char* 20	the valid release status of the model
	SADMRD	char* 6	release status date of the model
	SADMIC	char* 1	in change
	SADMVD	char* 6	valid until (date)
	SADREM	char* 70	remarks
	SADTIT	char* 40	title, description
	SADGFO	char* 20	geometry format
	SADDIM	char* 3	dimension
M	SADLGA	char* 1	local/global
	SADNAE	char* 3	number of assembly-elements
	SADPCO	char* 10	part_code
	SADMCO	char* 10	model_code
	SADUPG	char* 4	unified_parts_grouping
	SADMCL	char* 4	model_change_level
	SADCSP	char* 7	country_spec_package
	SADOEC	char* 7	optional_equipment_code
	SADEST	char* 8	engineering status
	SADULM	char* 4	unit_of_length_measurement
	SADPOD	char* 40	place_of_deposit
	SADDNR	char* 15	drawing_number
	SADDCI	char* 4	drawing_change_index
	SADSNR	char* 12	sheet_number

1)   M = mandatory

2)   P = (part of) primary key

The structure concerning administrative data and used in the program is:

SSFADA   char* 506

5.2.2 The Assembly Tree Table

This table contains the structure of each assembly, saying which elements comprise an assembly.

TABLE-NAME   : NI_ASSEMBLY_TREE

Type	Column	Data Type	Key	Description
M	SATFAT	char* 14	P	identification of the set of administrative data describing the assembly (=father)
M	SATSON	char* 14	P	identification of the set of administrative data that is an element of the assembly

The structure concerning assembly trees and used in the program is:

SSFATR   char* 28

5.2.3 The Closed Model Table

This table contains closed models, i.e., geometry in any format, part program and NC-data without any knowledge of their internal structure. The restriction of the length of a column in ORACLE (max. 240 chars) forces the splitting of the data into single blocks.

TABLE_NAME :   NI_CLOSED_MODEL

Type	Column	Data Type	Key	Description
M	SCMIDF	char*14	P	identification code
M	SCMANO	char*5	P	ascending number to be able to keep the origin sequence
M	SCMREC	char*5		number of records of CAD/CAM-data
M	SCMRCL	char*5		record length
M	SCMDA1	char*240		model data1
	SCMDA2	char*240		model data2
	SCMDA3	char*240		model data3
	SCMDA4	char*240		model data4
	SCMDA5	char*240		model data5

The structure concerning parts of closed models and used in the program is:

SSFCMO   char* 1229

5.2.4 The B_Rep Table

This table contains the values for the B_Reps:

TABLE_NAME :   NI_B_REP

Type	Column	Data Type	Key	Description
	SCOPE_NUMBER_EN	integer;		scope number of the enclosing entity
	B_REP_USER_NAME	char(80);		userdefined B_Rep name
	B_REP_INTERNAL_NAME	integer;		B_Rep internal name
	BR_SCOPE_NUMBER	integer;		B_Rep scope number

5.2.5 The Shell Table

Two tables contains the values for the shells. The first table is defined as:

TABLE_NAME   :   NI_SHELL

Type	Column	Data Type	Key	Description
	SCOPE_NUMBER_SH	integer;		scope number of the enclosing entity
	SHELL_NAME	integer;		shell name
	B_REP_INT_NAME	integer;		internal name of the according B_Rep. This is used to get the B_Rep_Result!
	SHELL_IDENTIFIER	integer;		identifier to build relations between the shell and its faces.

The second table is used to handle the faces (with their order) of the shells:

TABLE_NAME   :   NI_SHELL_HELP

Type	Column	Data Type	Key	Description
	SHELL_IDENTIFIER	integer;		identifier to build relations between the shell and its faces.
	SH_ORDER	integer;		order number of the face-element. This order number corresponds to the index used in the array's IARFAC and IAWFAC in the routines NIINSH, NISLSH and NIUPSH.
	SH_FACE_REF	integer;		name of the referenced face entity

5.2.6 The Face Table

For the purpose of normalization two tables are used to handle the face information. The first table is defined like the following:

TABLE_NAME   :   NI_FACE

Type	Column	Data Type	Key	Description
	SCOPE_NUMBER_FA	integer;		scope number of the enclosing entity
	FACE_NAME	integer;		face name
	PS_NAME	integer;		name of the referenced planar_surface
	FA_PS_ORIENT	logical;		orientation of the referenced planar_surface entity
	FACE_IDENTIFIER	integer;		identifier to build relations between the face and its loops.

The second table is used to handle the loops (with their order) of the faces:

TABLE_NAME   :   NI_FACE_HELP

Type	Column	Data Type	Key	Description
	FACE_IDENTIFIER	integer;		identifier to build relations between the face and its loops
	FA_LOOP_ORDER	integer;		order number of the loop-element. (this order number corresponds to the index used in the array's IARLOP and IAWLOP in the routines NIINFA, NISLFA and NIUPFA)
	FA_LOOP_REF	integer;		name of the referenced loop entity

5.2.7 The Loop Table

For the purpose of normalization two tables are used to handle the loop information. The first table is defined like the following:

TABLE_NAME   :   NI_LOOP

Type	Column	Data Type	Key	Description
	SCOPE_NUMBER_LP	integer;		scope number of the enclosing entity
	LOOP_NAME	integer;		loop name
	LOOP_IDENTIFIER	integer;		identifier to build relations between the loop and its edges.

The second table is used to handle the edges and their orientations (with their order) of the loops:

TABLE_NAME   :   NI_LOOP_HELP

Type	Column	Data Type	Key	Description
	LOOP_IDENTIFIER	integer;		identifier to build relations between the loop and its edges
	LP_EDGE_ORDER	integer;		order number of the loop-element. (this order number corresponds to the index used in the array's

			IAWEDR, IAREDR, LAREDO and LAWEDO in the Routines NIINLP, NISLLP and NIUPLP)
LP_EDGE_REF	integer;		name of the referenced edge entity
LP_EDGE_ORIENT	logical;		orientation of the referenced edge entity

5.2.8 The Edge Table

This table contains the values for the edges:

TABLE_NAME   :   NI_EDGE

Type	Column	Data Type	Key	Description
	SCOPE_NUMBER_ED	integer;		scope number of the enclosing entity
	EDGE_NAME	integer;		edge name
	ED_LINE	integer;		name of the line entity
	ED_START_VERTEX	integer;		name of the start_vertex entity
	ED_END_VERTEX	integer;		name of the end_vertex entity

5.2.9 The Vertex Table

This table contains the values for the vertices:

TABLE_NAME   :   NI_VERTEX

Type	Column	Data Type	Key	Description
	SCOPE_NUMBER_VE	integer;		scope number of the enclosing entity
	VERTEX_NAME	integer;		vertex name
	VE_POINT	integer;		point_name referenced

5.2.10 The Planar Surface Table

This table contains the values for the planar_surfaces:

TABLE_NAME   :   NI_PLANAR_SURFACE

Type	Column	Data Type	Key	Description
	SCOPE_NUMBER_PS	integer;		scope number of the enclosing entity
	PLANAR_SURFACE_NAME	integer;		planar_surface name
	PS_POINT	integer;		point on the planar_surface
	PS_DIRECTION	integer;		normal_direction of the planar_surface
	PS_REF_POINT	integer;		reference_point of the planar_surface

5.2.11 The Line Table

This table contains the values for the lines:

TABLE_NAME   :   NI_LINE

Type	Column	Data Type	Key	Description
	SCOPE_NUMBER_LI	integer;		scope number of the enclosing entity
	LINE_NAME	integer;		line name
	LI_POINT	integer;		point of the line
	LI_DIRECTION	integer;		direction of the line

5.2.12 The Point Table

This table contains the values for the points:

TABLE_NAME   :   NI_POINT

Type	Column	Data Type	Key	Description
	SCOPE_NUMBER_PT	integer;		scope number of the enclosing entity
	POINT_NAME	integer;		point name
	X-COMPONENT_PT	real;		x-component of the point
	Y-COMPONENT_PT	real;		y-component of the point
	Z-COMPONENT_PT	real;		z-component of the point

5.2.13 The Direction Table

This table contains the values for the directions:

TABLE_NAME   :   NI_DIRECTION_TABLE

Type	Column	Data Type	Key	Description
	SCOPE_NUMBER_DI	integer;		scope number of the enclosing entity
	DIRECTION_NAME	integer;		direction name
	X-COMPONENT_DI	real;		x-component of the direction
	Y-COMPONENT_DI	real;		y-component of the direction
	Z-COMPONENT_DI	real;		z-component of the direction

5.2.14 The Access Right Table

This table contains all access rights, saying for each set of administrative data who is allowed to read, write or delete it.

TABLE_NAME :   NI_ACCESS_RIGHTS

Type	Column	Data Type	Key	Description
M	SACIDF	char* 14	P	identification of the set of administrative data the access right belongs to
M	SACIUG	char* 20	P	groupid of the group to which the access right is assigned
M	SACRWD	char* 1		the kind of access (r=read, w=write, d=delete)
	SACIDA	char* 6		the date the access right is inserted
	SACTIM	char* 6		the time the access right is inserted
	SACUID	char* 20		the userid of the user who inserted the access right

The structure concerning access rights and used in the program is:

SSFACC   char* 67

5.2.15 The User/Group Definition Table

This table contains all users and groups existing in the system and to which access rights can be assigned.

TABLE_NAME :   NI_USER_GROUP_DEF

Type	Column	Data Type	Key	Description
M	SUSIUG	char*20	P	identification of user/group: userid or free chosen name for groups
M	SUSTYP	char*1		the type of the authority, to be able to distinguish between user, department as a group or any given group called project
	SUSUNA	char* 20		name of the user (empty in case of groups)

The structure concerning user/group definitions and used in the program is:

SSFUSG   char* 41

5.2.16 The Group Tree Table

This table contains the structure of all groups saying who is father or son of whom.

TABLE_NAME :   NI_GROUP_TREE

Type	Column	Data Type	Key	Description
M	STRFAT	char*20	P	identification of the group that is father of the user/group

| M | STRSON | char*20 | P | identification of the user or group that is son of the group |

The structure concerning a father-son relation and used in the program is:

    SSFTRE   char* 40

5.2.17 The Structure Definition Table

This table describes the internal structure of each string variable used in the program. It enables a dynamic handling of structures in the database interface at SQL-level. It is possible to change structures as they have no influence in this interface.

TABLE_NAME :   NI_STRUCTURE

Type	Column	Data Type	Key	Description
M	SSTSTC	char* 2	P	code of the kind of structure
M	SSTSFL	char* 3	P	name of the field contained in the structure
M	SSTOFF	char* 4		the offset of the field
M	SSTLEN	char* 4		the length of the field
	SSTKEY	char* 1		signs whether this field belongs to the primary key or not

The structure concerning structure definitions and used in the program is:

        SSFSTR   char* 18

Example: Description of the structure SSFACC = access rights

SSTSTC	SSTSFL	SSTOFF	SSTLEN	SSTKEY
AC	IDF	1	14	P
AC	GID	15	20	P
AC	RWD	35	1	P
AC	DAT	36	6	
AC	TIM	42	6	
AC	UID	48	20	

5.2.18 The Predefined SQL-Statement Table

This table contains all predefined SQL-statements or parts of SQL-statements which can be concatenated and cannot be generated as, e.g., the access control.

TABLE_NAME :   NI_SQL

Type	Column	Data Type	Key	Description
M	SSQFCO	char* 6	P	SQL function code, identifies an SQL-statement or a part of an SQL-statement
M	SSQSNU	char* 5	P	sequence number of the SQL-statement
M	SSQSTA	char* 70		SQL-statement

The structure concerning predefined SQL-statements and used in the program is:

     SSFSQL   char* 81

5.2.19 The Plausibility Check Table

To avoid frequent changes of the source code caused by plausibility checks being changed (especially administrative data), the rules are classified and inserted into database tables. For each class a standard method will be developed to check the relevant rules during the database manipulation (= at runtime).

Rules concerning single fields:

class	Rule
M	Mandatory
	The field is mandatory, i.e., it must have a value, may not be blank [1]
V	Set of given values
	The field must have a value of a given set of values or be empty.
C	Capital letters
	Each character of the field must be a capital letter or be empty.
S	Small letters
	Each character of the field must be a small letter or be empty.
F	Special formats
	The field must have a value in a special, predefined format or be blank (e.g. xxx.yy where x signifies letters and y signifies numbers).
C	Complete
	The field must be completely filled, each character must have a value.
L	Letters
	Each character of the field must be a letter or a blank.

The following classes are introduced to check datatypes, because all data necessary for administration are realized and implemented as characters. The reason is the aim to be as

---

[1] In this proposal there is no difference between a field that has no value and a field to which blanks are assigned. In both cases there will be no insert into the database. In ORACLE thie field is NULL.

independent as possible of the database management system used. Using only character types one is able to maintain a clear, uniform interface to the database.

D   Date
        The field must have a valid date in the form YYMMDD or be empty.

T   Time
        The field must have a valid time in the form HHMMSS or be empty.

N   Numeric
        Each character of the field must be a number or blank, but numbers may not be divided by blanks.

A   Alphanumeric
        Each character of the field must be a letter, a number or a blank.

By handling plausibility checks in this generic way, each company is able to define its own specific checks.

Rules concerning more than one field include implications.

A rule of the class implication may be, e.g.,

        "if the field USER_NAME has a value then the field USER_ID must have a value"

or

        "if the field TYPE_OF_MODEL has the value PART_GEOMETRY then the field GEOMETRY_FORMAT must have a value"

the if- and the then- conditions can each be a logic expression, a combination of logic AND or logic OR.

In a first step the following type of implication will be realized:

        IF   field a has a special value/does not have a special value
                 AND field b has a special value/does not have a special value
                 AND ...

        THEN   field x must have a special value/does not have a special value
                 AND field y must have a special value/does not have a special value
                 AND ...

The logic AND may also be substituted by the logic OR, but not mixed inside the if- or the then-clause.

The restriction to this special form of implication has the advantage that these implication rules can be handled in a standard way. Being inserted in a database table, a check of these rules can be generated. A change of the rules means no intervention in the source code, only an update in the database.

For all other combinations, special routines must be written.

Example

If TYPE_OF_MODEL is PART_GEOMETRY then GEOMETRY_FORMAT is a mandatory field and PART_CODE is a mandatory field.

By handling implication checks in this generic way, each company is able to define its own specific checks.

a)   Table to manage plausibility checks

This table lists all fields that must be checked and assigns each of them to one of the classes described above

TABLE_NAME :   NI_CHECKS

Type	Column	Data Type	Key	Description
M	SCHPRI	char* 3		priority of the check; by this number the sequence of the checks can be fixed
M	SCHFLD	char* 6	P	name of the field the value of which must be checked
M	SCHCLA	char* 5	P	the sign of the class of check

The structure concerning a plausibility check and used in the program is :

        SSFCHE   char* 14

Example:

      the plausibility checks
            creating date (SADADC) is mandatory (class M)
            host operating system (SADHOS) must have a value of a given set (class V)
            creating date (SADADC) must be a valid date (class D)
      would cause the following inserts

SCHPRI	SCHFLD	SCHCLA
1	SADADC	M
2	SADADC	D
3	SADHOS	V

b)   Table of sets of given values

This table contains all valid values of a field

TABLE_NAME :   NI_VALUES

Type	Column	Data Type	Key	Description
M	SVAFLD	char* 6	P	name of the field the valid values of which are listed
	SVAVAL	char* 50	P	value that is valid for the given field

The structure concerning valid values and used in the program is:

SSFVAL   char* 56

Remark:

The single rows of the table must be connected by a logical OR for each field, so that, e.g.,

SVAFLD	SVAVAL
SADDIM	2
SADDIM	2.5
SADDIM	3
SADHOS	VMS
SADHOS	MVS
SADHOS	VM/CMS
SADHOS	NOS
SADHOS	UNIX

reads   the field "dimension" of the administrative data can have the value 2 OR 2.5 OR 3;
AND
the field "host operating system" of the administrative data can have the value VMS
OR MVS OR VM/CMS OR NOS OR UNIX

c)   Table of special formats

This table contains the specification of special formats: for each character of a field there are
defined the allowed values

TABLE_NAME :   NI_FORMAT

Type	Column	Data Type	Key	Description
M	SFOIFO	char* 3	P	identification of the specified format
M	SFOBEG	char* 3	P	range (beginning) of the field a part of the format is specified
M	SFOEND	char* 3	P	range (end) of the field a part of the format is specified
	SFONOT	char* 1		signs whether the char(s) must have the value given in the column SFOVAL, SFONUM or SFOLET (i) or may not (blank)
	SFOCVA	char* 5		value, valid for the char (or each char of the given range)
	SFONUM	char* 1		signed by "x", the char (or each char of the given range) must be a number
	SFOLET	char* 1		signed by "x", the char (or each char of the given range) must be a letter

The structure concerning format specification and used in the program is:

SSFFOR   char* 17

Example:

the format F1 shall be specified as
'V.XXXYZZ' with   V   =   must be a letter
                   X   =   every sign, but no number
                   Y   =   only blank
                   Z   =   must be letter, number or blank

SFOIDF	SFOBEG	SFOEND	SFONOT	SFOVAL	SFONUM	SFOLET
F1	1	1				X
F1	2	2				
F1	3	5	X		X	
F1	6	6		BLANK		
F1	7	8		BLANK	X	X

d)   Table of implications

This table contains all rules that can be represented in the form of an implication as defined in chap.4 "Semantic integrity and logic of transaction" of Document No. 14.87.

TABLE_NAME :   NI_IMPLICATION

Type	Column	Data Type	Key	Description
M	SIMIIM	char* 3	P	identification of the specified implication
M	SIMSEQ	char* 3	P	sequence number of the implication
M	SIMITC	char* 4		signs whether the row means the if-clause or the then-clause of the implication
M	SIMFNA	char* 6		the name of the concerning field
	SIMNOT	char* 1		X means given the field may not have the value given in column SIMVAL means that the field must have the value
M	SIMVAL	char* 50		value the field must have or may not have

The structure concerning implications and used in the program is:

SSFIMP   char* 67

Example:

if type of model (SADTOM) is 'PART_GEOMETRY'
or type of model is 'TOOL_GEOMETRY'
then geometry format (SADGFO) is mandatory (not blank)
and part_code (SADPCO) is mandatory

SIMIIM	SIMSEQ	SIMITC	SIMFNA	SIMNOT	SIMVAL I
IQ1	1	IF	SADTOM		PART_GEOMETRY
IQ1	2	OR	SADTOM		TOOL_GEOMETRY
IQ1	3	THEN	SADGFO	X	BLANK
IQ1	4	AND	SADPCO	X	BLANK

5.2.20 The Error Tables

To handle the errors two tables are defined.

a)   Error/Message table

TABLE_NAME   :   NI_ERROR_MESSAGE

Type	Column	Data Type	Key	Description
	NR	integer;		Error number
	MESSAGE	CHAR(80)		Message text

b)   Error-kind table

TABLE_NAME   :   NI_ERROR_KIND

Type	Column	Data Type	Key	Description
	KIND	integer;		Error kind number
	MEANING	CHAR(80)		Error kind message

The following Table 5.1 shows the values used in the implementation.

Table 5.1 Error-kind table

Kind	Meaning
0	OK-STATUS
1	FATAL ERROR
2	WARNING
3	INFORMATION

In Table 5.2 the possible combinations between the the kind of error and the error-number (with its corresponding message-text) are shown. The three levels at which errors can happen are expressed by a short header of each message :

☞   "NI:" indicates an error in a neutral interface routine

☞   "DB:" indicates an error in the data base system

☞   "OS:" indicates an error in the operating system

Table 5.2 Possible errors and messages:

No.		Messages
0	NI:	routine was successfully completed
1	NI:	there is no opened scope
2	DB:	table full , can't insert
3	DB:	not enough space in the database, can't insert
4	NI:	real value out of range
5	NI:	integer value out of range
6	NI:	entity with this name already exists in the table
7	NI:	invalid name for an entity, only >0 allowed
8	NI:	name not found
9	NI:	name found in an outer scope
10	NI:	name out of range
11	NI:	number of selected entries greater than parameter for dimensioning array's
12	NI:	nothing to list
13	NI:	nothing to insert, number of elements <1
14	NI:	field overflow
15	NI:	nothing to update, number of elements <1
16	NI:	B_Rep name is empty
17	NI:	B_Rep name is too long
18	NI:	B_Rep name already exists
19	NI:	B_Rep name not found
20	NI:	unable to delete an opened B_Rep
21	NI:	another B_Rep is still open, please close first
22	NI:	No B_Rep opened to close
23	NI:	No B_Rep opened to list statistics
24	NI:	B_Rep is empty
25	NI:	username is empty
26	NI:	password is empty
27	NI:	remaining unused attributes have been deleted from the scope of the closed entity
28	NI:	assembly can't be inserted into this scope
29	NI:	assembly name is empty
30	NI:	assembly name is to long
31	NI:	assembly name already exists
33	NI:	unable to delete an opened assembly
34	NI:	assembly is empty
35	NI:	no assembly opened to list statistics
36	NI:	no assembly opened to close
101	NI:	creation values missing
102	NI:	creator must be an authorized user in the system
104	NI:	the input string is empty/identifier does not exist
105	NI:	no further elements because the identifier is not one of an assembly
107	NI:	the element to be included does not exist
109	NI:	the assembly does not possess the element

110	NI:	the excluded element was the last one of the assembly
111	NI:	set of the administrative data with the given identifier does not exist
113	NI:	the group/user does not exist in the system
114	NI:	the kind of access must be: R=read, W=write, D=delete
115	NI:	access right already exists
116	NI:	the given access right does not exist
117	NI:	no deletion because the access right is the last one of the set of administrative data
118	NI:	ascending number is empty
119	NI:	number of records is empty
120	NI:	there are no models to be inserted
121	NI:	closed model already exists
123	DB:	cursor does not exist
124	NI:	no (more) records selected
128	DB:	database system not active
130	DB:	record not found in the database
131	NI:	not allowed to change more than one record
132	NI:	invalid function code
133	NI:	input parameter is out of range
134	DB:	not enough bind variables for SQL-statement
135	DB:	invalid oracle userid/password
136	DB:	error during logon database
137	DB:	database error concerning the COMMIT command
138	DB:	database error concerning the ROLLBACK command
139	DB:	invalid cursor number
140	DB:	cursor already used by another selection
141	DB:	no cursor available
142	DB:	SQL buffer overflow
143	DB:	database error
144	DB:	database error
145	DB:	database error
146	DB:	buffer overflow
147	DB:	too many bind variables
148	DB:	duplicate insert trial in unique index
149	DB:	database error
150	DB:	select list is empty
151	DB:	too many select variables in the select list
152	DB:	error in function end select (close cursor)
153	DB:	cursor not open
154	DB:	error in function fetch
155	DB:	fatal internal index error
156	DB:	invalid value "blank" found
157	DB:	error in function "get date and time"
158	DB:	record in buffer not found
159	DB:	error internal read
160	DB:	invalid structure code
161	DB:	primary key field is empty
162	DB:	external SQL-statement must begin with "AND" or "ORDER BY"

163	OS:	error during FORTRAN read
164	OS:	system call: error during command execution
165	OS:	system call: command string is empty
166	DB:	too much storage demanded
167	DB:	invalid type of cursor
168	DB:	demanded storage is null
169	DB:	invalid record number
170	OS:	error during call for actual userid
197	NI:	circle found during include
198	NI:	local / global definition is not "G" or "L"
200	NI:	database name is empty
201	NI:	database password is empty
202	NI:	identifier of administrative data in access right is empty
203	NI:	user/group is empty
204	NI:	length of field condition string does not confirm with the number of fields of the corresponding value string
205	NI:	the identifier of access rights (= ID of data ID of user/group) or part of it is empty
206	NI:	CAD/CAM system name or CAD/CAM system release is empty
207	NI:	no unique identifier created
208	NI:	host-type is empty
209	NI:	host-operating-system or host-operating-system-release the model comes from is empty
210	NI:	type of model is empty
211	NI:	user is not allowed to delete administrative data set
212	NI:	warning no closed models deleted
213	NI:	identifier of assembly (owner) is empty
214	NI:	identifier of data set to be included is empty
215	NI:	local/global definition empty
216	NI:	user not allowed to include/update data set
217	NI:	element to be included already exists in an assembly
218	NI:	element already included in the assembly
219	NI:	there is no field to be changed
220	NI:	date of creation of administrative data set may not be changed
221	NI:	time of creation of administrative data set may not be changed
222	NI:	user-id of the creator of the administrative data set must not be changed
223	NI:	receiving date of administrative data set may not be changed
224	NI:	receiving time of administrative data set may not be changed
225	NI:	date the model was changed may not be changed
226	NI:	time the model was changed may not be changed
227	NI:	user who changed model must not be changed
228	NI:	date of changing adm-data-set may not be changed
229	NI:	time of changing adm-data-set may not be changed
230	NI:	user who changed adm-data-set may not be changed
231	NI:	CAD/CAM-system name may not be changed
232	NI:	number of element in an assembly may not be changed
233	NI:	user is not allowed to select the set of administrative datawith the given identifier

234	NI:	administrative data set is still used in other assemblies
235	NI:	user is not allowed to insert access right
236	NI:	user is not allowed to delete access right
237	NI:	there is no access right "DELETE" for the identifier in the system
238	NI:	user is not allowed to update access right
239	NI:	no update because the access right with the given identifier is the last one with "DELETE" for this administrative data set
241	NI:	local sets of administrative data without being referenced by an assembly
242	NI:	part code may not be blank
243	NI:	model code may not be blank
244	NI:	model change level may not be blank
245	NI:	engineering status may not be blank
246	NI:	bottom of assembly tree already reached
250	NI:	type of user/group may be "U"or "G"
251	NI:	name of user may not be blank
252	NI:	userid/groupid may not be blank
999	NI:	error during selection of the error message

6   NIDIC - The Neutral Interface Data Dictionary

The main task of a data dictionary is to administrate data fields and their relations as data structures or data base tables. It is a data definition catalogue.

Being able to represent aggregations and references between entities, a data dictionary can also be used to support the work of programming directly. For that purpose program generators and precompilers are necessary to generate code in a higher level programming language. Such a code can,e.g., be the whole set of definitions of local, global variables, input/output parameters used by a module. These generated parts of codes will be automatically linked to the program code during the compilation process. The essential profit compared with the so-called INCLUDE- and COPY-libraries is that one is able to connect administration and programming. This means that by using a data dictionary it is not only possible to recognize the effects of modifications but also to enjoy the great security of doing the changes automatically.

The following figure 6.1 describes the data model of neutral interface data directory.

The data and their relationships stored in the data dictionary can be used for

a) Information and design control

During system analyses and realization informations about structuring and referencing is necessary. The most important information for this step are:

- data catalogue
- data structure catalogue
- program modules: definitions and overviews
- calling hierarchy

In NIDIC it is possible to generate reports to show coherences and give overviews. The following examples show the report of a module description and a calling hierarchy.

Example :   Module Description

NIDLAD delete administrative data set

      Version: VAX/VMS

      Revision: 1.0

The subroutine deletes the set of administrative data identified by IDF. In addition all access rights and the model belonging to this set of administrative data will be deleted. In case of an assembly, the whole structure with all sets of administrative data, access rights and models will be deleted.

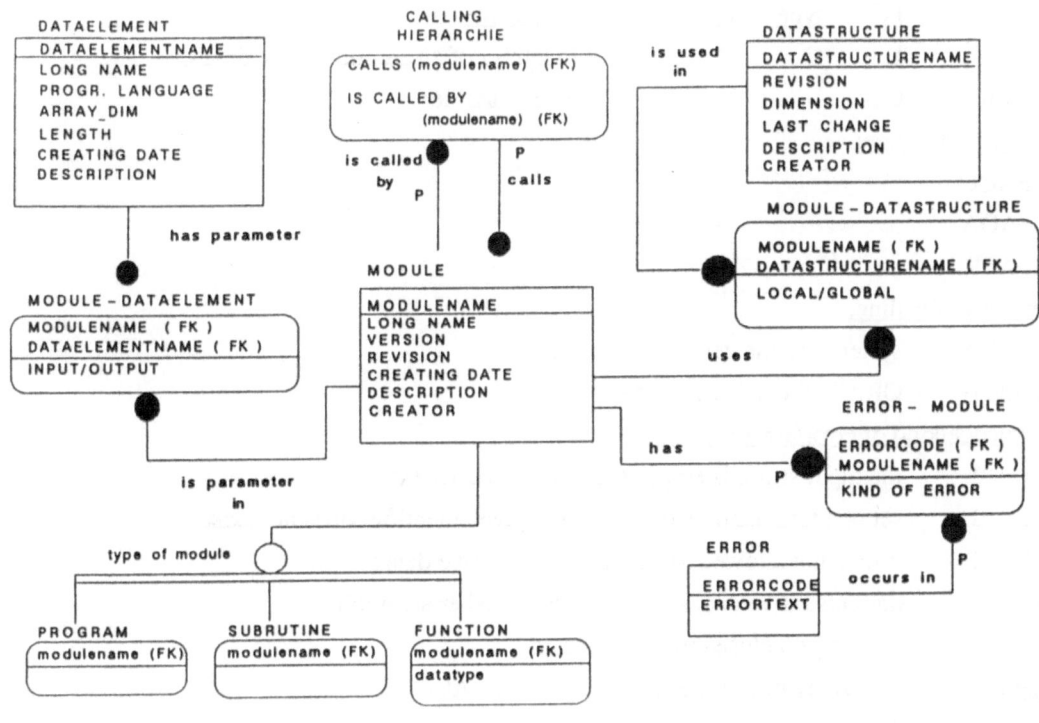

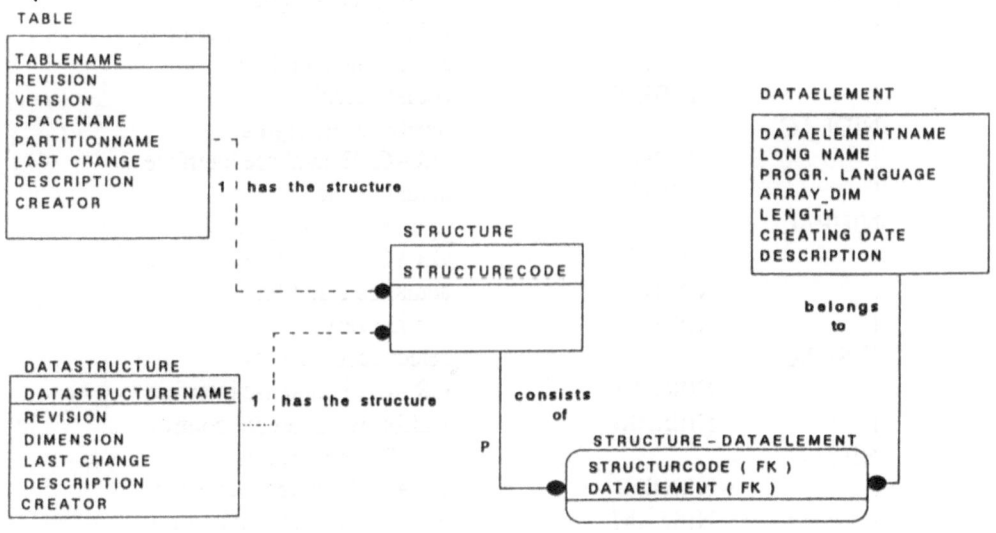

Fig. 6.1 The Data Model of NIDIC

Parameter :

name	I/O	type	length	long name
PIDF	I	C	14	identifier of an administrative dataset
IAWER	O	I	3	error variable

used data structure:

SLFACC

SLFADA

SLFATR

hierarchy of calling:

NISBUN	system routine getting the userid
NISODB	ORACLE database interface

returncode/kind of error/error message

104	1	the input string is empty / identifier deosn't exist
111	1	set of administrative data with the given identifier does not exist
211	1	user is not allowed to delete administrative data set
234	1	administrative data set is still used in other assemblies
212	1	warning no closed models deleted

Example: Hierarchy of routines called by routine ACCESS

Version: VAX/VMS

L0	L1	L2	L3
ACCESS			ACCESS right function
	NIINAC		insert access rights
	!	NISODB	ORACLE database interface
	!	NISDZN	actual date and time
	!	NISBUN	actual userid
	NIDLAC		delete access rights
	!	NISODB	ORACLE database interface
	!	NISBUN	actual userid
	NIUPAC		update access rights
	!	NISODB	ORACLE database interface
	!	NISDZN	actual date and time
	!	NISBUN	actual userid
	NISLAC		select access rights
	!	NISODB	ORACLE database interface
	!	NISSBM	Buffer management control
	NILIAC		list all users/groups
	!	NISODB	ORACLE database interface
	!	NISSBM	Buffer management control
	NICTFC		rollback a transaction
	!	NISODB	ORACLE database interface
	NIREAD		fetch next record
	!	NISSBM	Buffer management control

NIENFC		end of selection
!	NISSBM	Buffer management control
NISLER		selection of error messages
!	NISODB	ORACLE database interface

b) Generating program code

To guarantee data consistency during implementation, the tool for generating parts of programs using information from the data dictionary is very helpful.

Using this tool it is possible to generate a

header   for each module containing the necessary data declarations, the history a description, version and more (see example)

call interface   with all defined parameters and parameter declarations

data base table   in the form of SQL-statements for creating tables, views, indexes, with all internal declarations like space initial, etc.

Example. Header of Routine NISLAC

```
-- FORTRAN FORMAT CODING NONUM
-- <<NISLAC>> $SUBPROGRAM ( IDF , PCURS , IAWER )
-- select access rights
-- +----------------------------------------------------------+
-- ! SYSTEM           : ESPRIT
-- ! REVISION         : 1.0
-- ! VERSION          : *
-- ! CREATOR          : FELDMEIER
-- ! DATE OF CREATION : 29-MAR-88
-- +----------------------------------------------------------+
-- ! DESCRIPTION OF THE FUNCTION:
-- ! Selection of all access rights belonging to the given set
-- ! of administrative data. This identifier may be the result
-- ! of the select routine; via the cursor, which is an output
-- ! of this routine, one is able to fetch the selected sets of
-- ! access rights from the buffer one after the other by
-- ! the procedure FETCH
-- +----------------------------------------------------------+
-- ! COMMONs   :
-- +----------------------------------------------------------+
-- ! FILES     :
-- +----------------------------------------------------------+
-- ! CALLED MODULES  :
-- ! MODULE    : NAME
-- ! NISODB    : ORACLE database interface
-- ! NISSBM    : Buffer Management Control
-- +----------------------------------------------------------+
-- ! CHANGES   :
-- ! 12-DEC-88   Feldmeier test : only user with access right
-- !             READ are allowed to select access rights
-- +----------------------------------------------------------+
--
```

```
--    declaration of constants
--
C     Max. number of free blocks in the buffer
      INTEGER WBBLCK
      PARAMETER ( WBBLCK = 100 )
C
C     Max. number of cursor for Buffer Manager
      INTEGER WBCURS
      PARAMETER ( WBCURS = 100 )
C
C     Max. number of records of each cursor for Buffer Manager
      INTEGER WBRCM
      PARAMETER ( WBRCM = 100 )
C
C     offset system buffer
      INTEGER WSBOFF
      PARAMETER ( WSBOFF = 25500 )
C
C     unit number of the output device
      INTEGER WUNR
      PARAMETER ( WUNR = 5 )
C
C     dimension of data element IAWER
      INTEGER XIAWER
      PARAMETER ( XIAWER = 3 )
C  --
  $PARAMETER
    $INPUT
C
C                         administrative data identifier
      CHARACTER * ( * )   IDF
    $INOUT
    $OUTPUT
C                         logical cursor number
      INTEGER             PCURS
C
C                         error variable
      INTEGER             IAWER (3)
C  --
$LOCAL
C
C                         administrative data identifier
      CHARACTER * 14      SACIDF
C
C                         identification of user/group
      CHARACTER * 20      SACIUG
C
C                         kind of access
      CHARACTER * 1       SACRWD
C
C                         data access right is inserted
```

```
      CHARACTER * 6        SACIDA
C
C                          time access right is inserted
      CHARACTER * 6        SACTIM
C
C                          userid of user who inserted  access right
      CHARACTER * 6        SACUID
C
      CHARACTER * 67       SLFACC
      EQUIVALENCE ( SACIDF , SLFACC ( 1 : 14 ) )
      EQUIVALENCE ( SACIUG , SLFACC ( 15: 34 ) )
      EQUIVALENCE ( SACRWD , SLFACC ( 35: 35 ) )
      EQUIVALENCE ( SACIDA , SLFACC ( 36: 41 ) )
      EQUIVALENCE ( SACTIM , SLFACC ( 42: 47 ) )
      EQUIVALENCE ( SACUID , SLFACC ( 48: 67 ) )
C
$BEGIN
  --
  -- ----------------------------------------------------------- -
  --              include module program code                    -
  -- ----------------------------------------------------------- -
  --
$RETURN
$ENDSUBPROGRAM <<NISLAC>>
```

c) Administration of programs

The administration of large systems can be supported by a data dictionary managing different versions and revisions, which means an automatic link of all modules called by a changed module.

7 Administrative Data Manager

7.1 Definition of the Screens

In this chapter, the different screens used to show interactively the functionality of the CAD*I Neutral Database Interface are described.

The Administrative Data Manager is a dialogue system by which a CAD/CAM constructor is able to inform her/himself interactively about existing models, versions and variants, and to exchange models with other CAD/CAM systems.

E0 Screen to get in

```
ESPRIT --- CAD*I --- WG4 --- NEUTRAL INTERFACE
       Test of neutral interface routines


    EEEEE SSSSS PPPPP RRRRR IIIII TTTTT
    E     S     P   P R   R   I     T
    EEEEE SSSSS PPPPP RRRRR   I     T
    E         S P     R R     I     T
    EEEEE SSSSS P     R   R IIIII   T



Command ===>
```

E1 Select Functions

```
ESPRIT --- CAD*I --- WG4 --- NEUTRAL INTERFACE
       Test of neutral interface routines

  0  stop test
  1  logon data base
  2  functions for sets of administrative data
  3  functions for access rights
  4  functions for closed models
  5  end of transaction
  6  interrupt with rollback
  7..functions for user/group definitions (DBA only)

  Command ===>
```

E2 Listing of Unreferenced Local Sets

```
ESPRIT --- CAD*I --- WG4 --- NEUTRAL INTERFACE
       Test of neutral interface routines

There are still unreferenced local sets
of administrative data

their identifiers are:


       _____
       _____
       _____
       _____


Include local element into an assembly before
end of transaction!

Command ===>
```

ADM0 Functions for Sets of Administrative Data

```
ESPRIT --- CAD*I --- WG4 --- NEUTRAL INTERFACE
       Test of administrative data routines

0   quit
1   insert set of administrative data
2   select set of administrative data
3   list elements of an assembly at the
    next lower level
4   include a new element into an assembly
5   exclude an element of an assembly
6   update set of administrative data
7   delete set of administrative data


Command ===>
```

ADM1 Set of Administrative Data (page 1)

```
ESPRIT --- CAD*I --- WG4 --- NEUTRAL INTERFACE
       set of administrative data

title: _____

drawing_no          : _____ part_code        : _____
drawing_change      : _____ model_code       : _____
model_change_level: _____ unit.parts_group: _____

geometry_format: _____ dim: ___

variant: _____ in_change: __
version: _____

remarks: _____


Command ===>
```

ADM1 Set of Administrative Data (page 2)

```
ESPRIT --- CAD*I --- WG4 --- NEUTRAL INTERFACE
         Set of administrative data

CAD/CAM_SYSTEM                    RECEPTION
 name    : _____          date: _____
 release: _____           time: _____

HOST                             CREATION
 type: _____          date: _____
 operating_system: _____         time: _____
 os_release: _____           user: _____

MODEL_DATA_CHANGING              ADM_DATA_CHANGING
 date: _____                   date: _____
 time: _____                   time: _____
 user: _____               user: _____


 Command ===>
```

ADM2 Elements of an Assembly

```
ESPRIT --- CAD*I --- WG4 --- NEUTRAL INTERFACE
       Test of administrative data routines

the elements at the next lower level of the
assembly with the identifier _____ are:

identifiers:
            _____
            _____
            _____
            _____
            _____


Command ===>
```

ADM3 List of Sets of Administrative Data

```
ESPRIT --- CAD*I --- WG4 --- NEUTRAL INTERFACE
     selection of sets of administrative data

S  IDENTIFIER  TITLE      CREATOR C/C-SYS  FORMAT
_  _____ _____ _____ _____ _____
_  _____ _____ _____ _____ _____
_  _____ _____ _____ _____ _____
_  _____ _____ _____ _____ _____
_  _____ _____ _____ _____ _____


Command ===>
```

ADM4 Include/Exclude Assembly Elements

```
ESPRIT --- CAD*I --- WG4 --- NEUTRAL INTERFACE
   include / exclude assembly elements

   identifier of the assembly: _____

   identifier(s) of the element(s)
   to be included/excluded:     _____
                                _____
                                _____
                                _____


Command ===>
```

DM5 start processor

```
ESPRIT --- CAD*I --- WG4 --- NEUTRAL INTERFACE
      Test of access right routines

      1  start processor STRIM
      2  start processor DICAD
      3  start processor NEUTRAL FILE FORMAT


FILE_SPECIFICATION: _____


   Command ===>
```

ACC0 Functions for Access Rights

```
ESPRIT --- CAD*I --- WG4 --- NEUTRAL INTERFACE
      Test of access right routines

      0  quit
      1  insert access rights
      2  select access rights
      3  update access rights
      4  delete access rights
      5  list all users/groups


   Command ===>
```

ACC1 Access Rights

```
ESPRIT --- CAD*I --- WG4 --- NEUTRAL INTERFACE
            access rights

   IDENTIFIER: _____

   GROUP-/USER-ID                ACCESS-RIGHT


  _  _____            ___
  _  _____            ___
  _  _____            ___
  _  _____            ___

   Command ===>
```

ACC2 List of Users/Groups

```
ESPRIT --- CAD*I --- WG4 --- NEUTRAL INTERFACE
          list of users / groups


   GROUP: _____

   list of users/groups at the next lower level:


                  _____
                  _____
                  _____
                  _____
                  _____


   Command ===>
```

CMO0 Functions for Closed Models

```
ESPRIT --- CAD*I --- WG4 --- NEUTRAL INTERFACE
        Test of closed model routines



         0   quit
         1   insert closed model
         2   select closed model
         3   delete closed model


   Command ===>
```

CMO1 Closed Model

```
ESPRIT --- CAD*I --- WG4 --- NEUTRAL INTERFACE
              closed model


   IDENTIFIER: _____


   FILE_SPECIFICATION: _____
                       _____


   Command ===>
```

7.2 Dialogue Screen Sequences

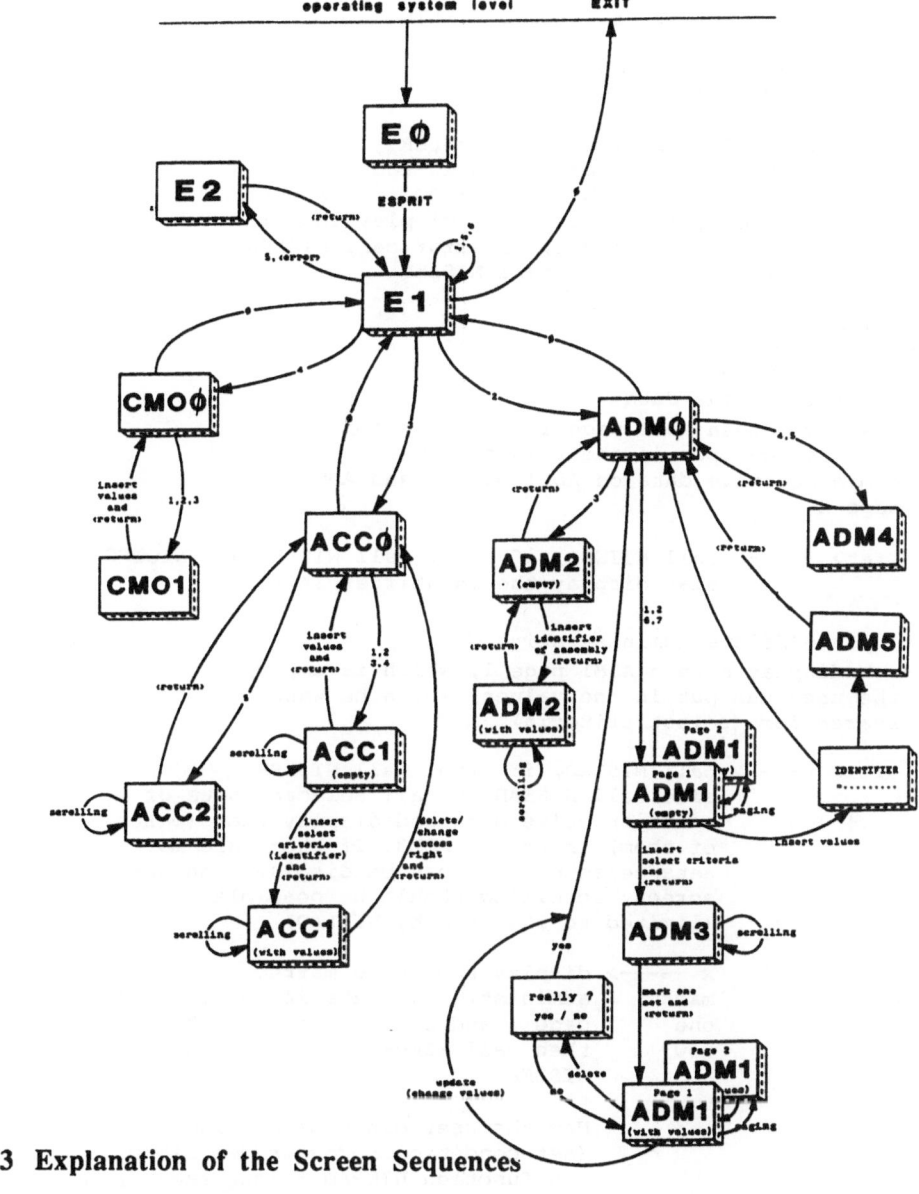

7.3 Explanation of the Screen Sequences

This describes which input of the user will call which neutral interface routine.

The input ESPRIT at operating system level displays the screen E0 to get in, and by <ret> the screen E1 to choose the function.

SCREEN E1

```
the input          will cause
0   ----->  call NICLWO, if error call NISLER
1   ----->  call NIOPWO
2   ----->  display screen ADM0
3   ----->  display screen ACC0
4   ----->  display CMO0
5   ----->  call NIETFC, if error call NISLER
            if there are still unreferenced local sets
            (ISWLOC >= 0) call NIREAD(ISWLOC,...) for
            all selected identifiers and display them in
            screen E2; call NIENFC, if error call NISLER
6   ----->  call NICTFC, if error call NISLER
```

SCREEN ADM0

```
0 -----> display screen E1

a) insert administrative data
1 -----> display screen ADM1 page 1, which is empty;
         the user can put in the values of the set of
         administrative data on ADM1 page 1 and ADM1
         page 2

         <ret> -----> call NIINAD, if error call NISLER and NICTFC
                      else display the identifier SADIDF


b) select/update/delete administrative data
2/6/7 -----> display screen ADM1 page 1, which is empty;
             the user can put in the values, which he wants to
             search for (select criteria)

         <ret> -----> call NISLAD, if error call NISLER
                      else call NIREAD for all selected sets of
                      administrative data and display them (parts
                      of them) in screen ADM3. If there are more
                      sets selected than can be displayed on one
                      screen, scrolling should be possible
                      (limited to 100 sets by NISLAD)

                      x -----> display the whole marked set of
                      mark     administrative data in screen ADM1
                      one      page 1 and 2 by calling NIREAD;
                      row      then call NIENFC, if error call
                               NISLER.

                               Now the user can update values
                               (set condition codes as described
                                in function NIUPAD - look remark)
                               <upd> -----> call NIUPAD, if error
                                            call NISLER and NICTFC

                               or delete the whole set
                               <del> -----> after verifying call
                                            NIDLAD, if error call
                                            NISLER and NICTFC


c) list elements of an assembly
3 -----> display screen ADM2;
         the user has to put in the identifier of the assembly
```

```
              <ret> -----> call NILIAD, if error call NISLER
                           else call NIREAD for all selected
                           identifiers and display them in screen ADM2;
                           scrolling should be possible; call NIENFC,
                           if error call NISLER
```

d) include/exclude elements of an assembly
```
4/5 -----> display screen ADM4
           the user has to put in the identifier of the assembly
           and the identifier(s) of the element(s) to be included
           or excluded; for each element to be included call
           NIICAD, to be excluded call NIECAD, if error call
           NISLER and NICTFC
```

Screen ACC0

```
0 ----->  display screen E1
```

a) insert access right
```
1 -----> display screen ACC1, which is empty;
         the user can put in the values: the identifier of the
         set of administrative data, the access right should
         belong to and the access rights (scrolling is possible)

         <ret> -----> for each access right call NIINAC, if error
                      call NISLER and NICTFC
```

b) select/update/delete access right
```
2/3/4 -----> display screen ACC1, which is empty;
         the user has to put in the identifier of the set of
         the set of administrative data of which he wants to
         select the belonging access rights

         <ret> -----> call NISLAC, if error call NISLER
                      else call NIREAD for each selected access
                      right, display them in screen ACC1
                      (scrolling is possible) and call NIENFC,
                      if error call NISLER.

                      Now the user can update access rights:
                      <upd> -----> call NIUPAC for each changed
                                   access right (only the kind of
                                   access   can be changed,
                                   NIUPAC checks this), if error
                                   call NISLER and NICTFC

                      or delete access rights
                      x -----> for each marked row call NIDLAC,
                      mark     if error call NISLER and NICTFC
                      rows
```

c) list all users/groups
```
5 -----> display screen ACC2
         the user has to put in the group-id of which he wants to
         get all users/groups at the next lower level

         <ret> -----> call NILIAC, if error call NISLER,
                      else for each selected user/group call
```

NIREAD and display them in screen ACC2;
call NIENFC, if error call NISLER

Screen CM00

0 -----> display screen E1

a) insert closed model
1 -----> display screen CM01,
         the user has to put in the identifier of the set of
         administrative data that describes the geometry and to
         put in the exact specification of the file the geometry
         is stored in

         <ret> -----> call NIINCM, if error call NISLER and NICTFC

b) select closed model
2 -----> display screen CM01,
         the user has to put in the identifier of the set of
         administrative data that describes the geometry, that
         shall be selected, and to put in the exact specification
         of the file, the geometry shall be written into

         <ret> -----> call NISLCM, if error call NISLER

c) delete closed model
3 -----> display screen CM01,
         the user has to put in the identifier of the set of
         administrative data that describes the geometry to be
         deleted

         <ret> -----> call NIDLCM, if error call NISLER and NICTFC

Remarks:

Set condition codes for each field that shall be changed:
    SCFADA = '0' (initialize SCFADA with 0)

now e.g. the field CAD/CAM-SYSTEM-NAME shall be updated (it is the second
field in the set of administrative data - look at Implementation
Guideline), then
    SCFADA(2:2) = '1'.

8   CAD*I Database Group Achievements and Experiences

8.1   CAD*I Database Processor Development

According to the developments performed, the situation reached in the CAD*I Database group is the following :

Two main families of CAD systems are available (see Fig. 8.1):

CAD systems (type 1) accessing data stored in the CAD*I Neutral file according to the CAD*I format (TECHNOVISION, PROREN),

CAD systems (type 2) accessing data stored in the CAD*I-ORACLE database according to the subset of the CAD*I database schema implemented by the CAD*I Database group (STRIM, TESTBED* , DICAD**)

Figure 8.1 details the CAD*I situation reached and the exchange mechanisms available between different CAD systems.

Indeed, with the interfaces developed by the CAD*I Database group, any CAD system is able to communicate with any other CAD system, to exchange CAD*I data according to one of the following mechanisms:

a)   Transfer between two CAD systems belonging to type 1 :

Transfer from the native format of the first CAD system to the CAD*I-Neutral file and transfer to the native format of the second CAD system (for example exchange between TECHNOVISION and PROREN),

b)   Transfer between two CAD systems belonging to type 2 :

Transfer from the native format of the first CAD system to the CAD*I-ORACLE database, creation of one ORACLE-Export file, and transfer to the native format of the receiving second CAD system (for example exchange between STRIM, DICAD and TESTBED),

c)   Transfer between one CAD system belonging to type 1 and one CAD system belonging to type 2 :

---

* TESTBED -   Geometric Modelling System from CAM-i developed by Cranfield Institute of Technology

** DICAD -   Dialog-oriented CAD system developed at RPK, University of Karlsruhe

Transfer from the native format of the first CAD system to the CAD*I-NEUTRAL FILE, transfer to the CAD*I-ORACLE database then transfer to the native format of the second CAD system (for example exchange between PROREN and STRIM),

Transfer from the native format of the first CAD system to the CAD*I-ORACLE database, transfer to the CAD*I-NEUTRAL FILE then transfer to the native format of the second CAD system (for example exchange between TESTBED and TECHNOVISION)

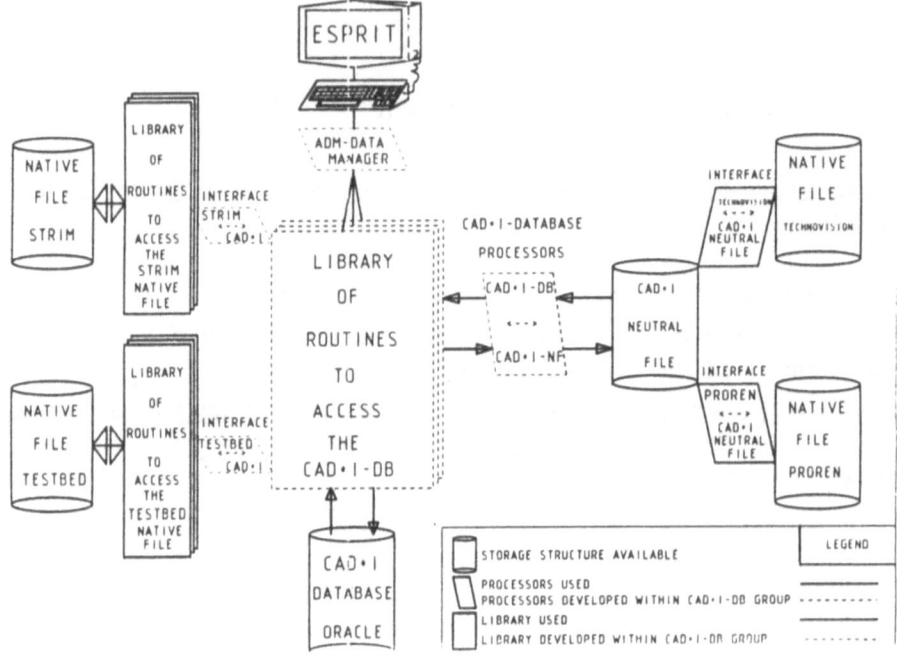

Fig. 8.1   CAD*I configuration available in the CAD*I Database group

In order to perform relevant tests on the interfaces developed in the CAD*I Database group, a workplan was defined to allow every partner concerned by a software implementation to test his/her own part.

The interfaces involved in this mechanism are described below.

8.1.1 CAD*I-ORACLE <--> STRIM 100 Interface

To transfer the whole structure of a B_REP from the CAD*I-ORACLE database to the STRIM100 CAD system and similarly in the other direction, the CAD*I Database group developed an interface software based on:

-    The CAD*I DATABASE INTERFACE routines mentioned above and especially the routines to select or insert entities in the CAD*I-ORACLE database,

-    The routines to write or read a VOLUME (equivalent to B_REP in CAD*I) in the STRIM 100 working area according to its own format (native format),

-    The routines to replace a polyhedric planar face associated with a complex patch by the same planar face bound to a planar surface supporting the face.

The algorithm includes the following steps:

- Reading TOP-DOWN in the CAD*I-ORACLE database or similarly in the STRIM working area according to the structure of the CAD*I format or similarly the Native format (STRIM)

- As soon as one topological entity is well defined (e.g. given all the geometrical and topological information describing this entity), the transfer of this entity is carried out and its address is passed to the above topological entity.

- The transfer is performed according to a loop algorithm and in order to define the volume completely.

The main points the algorithm takes into account carefully are:

- Non-duplication of the edges already transferred to the interface from the CAD*I database to the STRIM working area.

- Non-duplication of the vertices already transferred to the interface from the STRIM working area to the CAD*I database.

About ten models produced with the STRIM CAD system are available according to the CAD*I-ORACLE database format and were distributed to the partners of the CAD*I Database group to allow them to test their interfaces.

8.1.2 CAD*I NEUTRAL FILE <--> CAD*I-ORACLE Interface

This interface takes into account only B_REP models with planar surfaces, and is based on :

- The CAD*I database interface subroutines implemented by the CAD*I Database group and especially the routines to Read and Write CAD*I data in the Relational Database System ORACLE,

- The routines to Read or Write in the CAD*I Neutral file any CAD*I model stored according to the specifications of the CAD*I Neutral file for CAD geometry Version 3.3.

The Interface from CAD*I the Neutral file to the CAD*I-ORACLE database is composed of a CAD*I Database postprocessor written in standard FORTRAN-77 and working both in interactive and in batch mode. The CAD*I Database interface routines are used to insert in the CAD*I-ORACLE database the entities read in the Neutral file.

The postprocessor is built up by different modules, each of which handling a single entity or a special function.

The procedure to transfer a B_REP model from the CAD*I Neutral file into the CAD*I-ORACLE Database is the following:

- Generate a B_REP model(which contains only planar surfaces with straight edges, and lines and points as geometric entities) on CAD-system,

- Preprocess this model via the system dependent CAD*I-preprocessor into the Neutral file according to the CAD*I format,

- Start the CAD*I scanner/parser software to scan and parse the Neutral file,

- Start the CAD*I Database postprocessor to transfer the scanned data into the CAD*I-ORACLE database. Only B_REP model information of the CAD*I scanner/parser output is used.

Similarly, the interface from CAD*I-ORACLE database to the CAD*I Neutral file is composed of a database-preprocessor written in standard FORTRAN-77 and working in interactive mode. The CAD*I DATABASE INTERFACE routines are used to read in the CAD*I-ORACLE database the entities to be transferred in the CAD*I Neutral file.

8.1.3 AIS <--> CAD*I-ORACLE Interface

To improve the system modularity, to enable an independent development of processors and to ensure software exchangeability, a development based on a neutral programming interface is required. CAM-I's Geometric Modeling Program (GMP) has developed the Application Interface Specification (AIS), a neutral programming interface for geometric modeling.

At both systems the DICAD and the TESTBED system provide a programming interface based on the AIS. A new approach for the CAD*I database processor development was chosen. The processor based on

- The CAD*I Database Interface routines to read or write CAD*I entities in the ORACLE database,

- The Application Interface routines (AIS) based on the CAM-I AIS to access the native format of the TESTBED modeler,

- Euler operators to create a consistent B_REP model.

The Postprocessor Algorithm is based on the utilization of the CAD*I database interface routines to read CAD*I entities in the database and Euler operators to create B_REP models in the TESTBED native format.

The sequence of Euler operators to construct an object is not unique. A strategy had to be chosen that would be as fast as possible, would test as far as possible that the source structure is correct, and would not forget anything.

About ten models produced with the TESTBED and DICAD CAD system are available according to the CAD*I-ORACLE database format and were distributed to the CAD*I Database partners to allow them to test their interfaces.

8.2 Description of the Administrative Data Manager

In our demonstration program the work with the CAD*I Database starts with the insertion of administrative data into the database. After selecting the appropriate function on the menu screen the user is asked to fill in his/her own data in the subsequent panels. Here, first plausibility checks are done by the system.

In the following step the required processor is chosen. There exist processors for data with

- STRIM 100 format

- TESTBED format

-   neutral file format

    (according the specification of CAD*I Working Group 2)

With the choice of the processor, the file is specified, which contains the CAD/CAM data that should be stored in the CAD*I Database data base.

The processors receive the file name and file type of the CAD/CAM data and are implemented on the same host computer as the data base and the WG4 neutral interface. By calling routines from the neutral interface the CAD/CAM data are mapped onto the data structure of the data base.

The connection to the administrative data already inserted is built by the unique identifier created by the system.

Implemented in the project is only a subset of the geometric elements, which are created by various CAD/CAM systems and the various neutral file processors of CAD*I Working Group 2 respectively.

There are two possibilities to store CAD/CAM data in the data base. The first one is to decompose it in its single geometric elements and store these in various data structures. The second one stores the CAD/CAM data as a "black box" in the data base. This possibility is mostly used when the CAD/CAM data are to be exchanged between different CAD/CAM systems.

To receive CAD/CAM data from the WG4 data base the first step is a selection of administrative data. The user fills in his criterion in the appropriate panel. The system determines all the administrative data records according to the selection criterion. If we assume that the user is allowed to see all elements that have qualified, he takes a second selection to get a specific one from the result set. This is necessary because the result set contains in general more than one record. Then the chosen processor reads the CAD/CAM data using the routines from the neutral interface and converts it to the specific format of the receiving system.

It is also possible to create a neutral file according to the specification of CAD*I WG2.

The other functions of the implemented software allow the user to update or delete administrative data or to maintain the structure of assemblies.

If the user selects the line "functions for access rights" in the second panel after starting the dialogue, he can choose between the following operations:

-   insert access rights
-   select access rights
-   update access rights
-   delete access rights
-   list all users/groups

The last function allows the user to maintain the userids that belong to a user group, for example an organization unit. So it is possible to give an access right to the organization unit, and every member of the unit known to the system will have this given right.

Another use of the CAD*I Database and the neutral interface is the selection of single geometric elements from the data base. In the implemented version, however, the user has to know the names of the required elements or some other references to them. There is no support of more complex, geometry-oriented selection criteria.

8.3 Performed Intersection and Cycle Tests

Figures 8.2 and 8.3 show the result (hard copies) of the cycle tests performed between CAD systems DICAD, TESTBED and STRIM.

Figure 8.4 shows the transfer of a STRIM model with about 600 faces with the CAD*I-ORACLE <--> STRIM 100 interface. An Oracle export file has been sent to all the partners of the CAD*I database group to allow them to transfer this model according to their own interfaces.

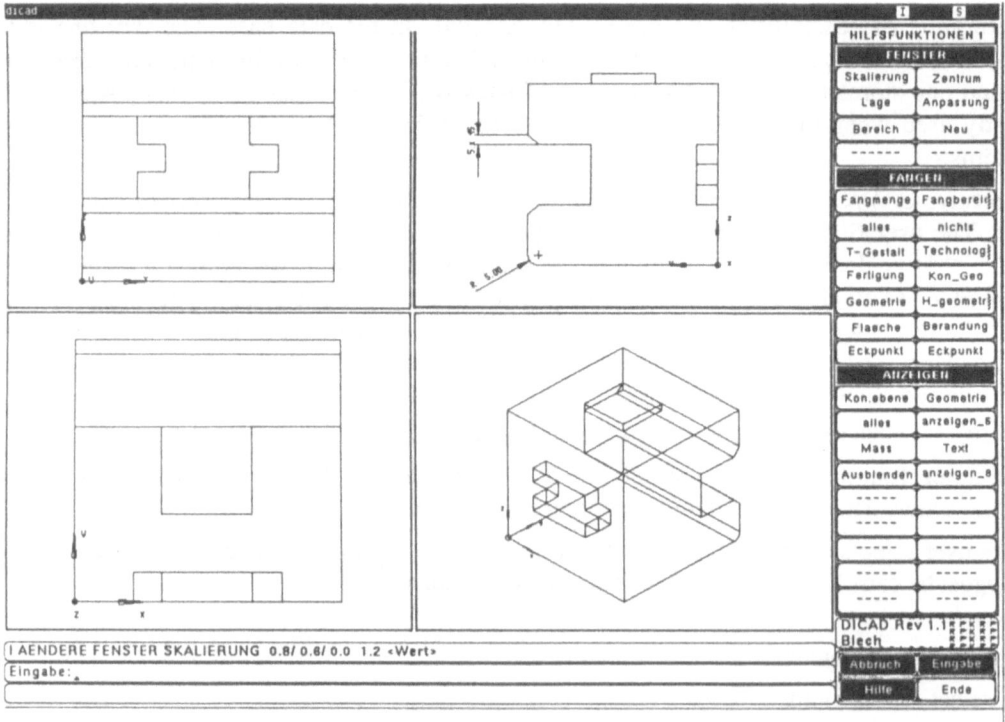

Fig. 8.2   Test part according to the DICAD system

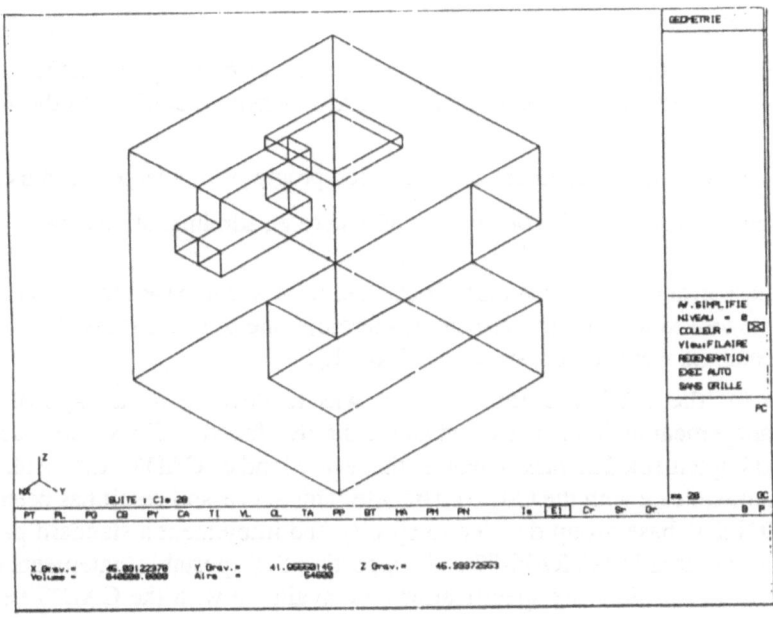

Fig. 8.3 Test part according to the STRIM CAD system

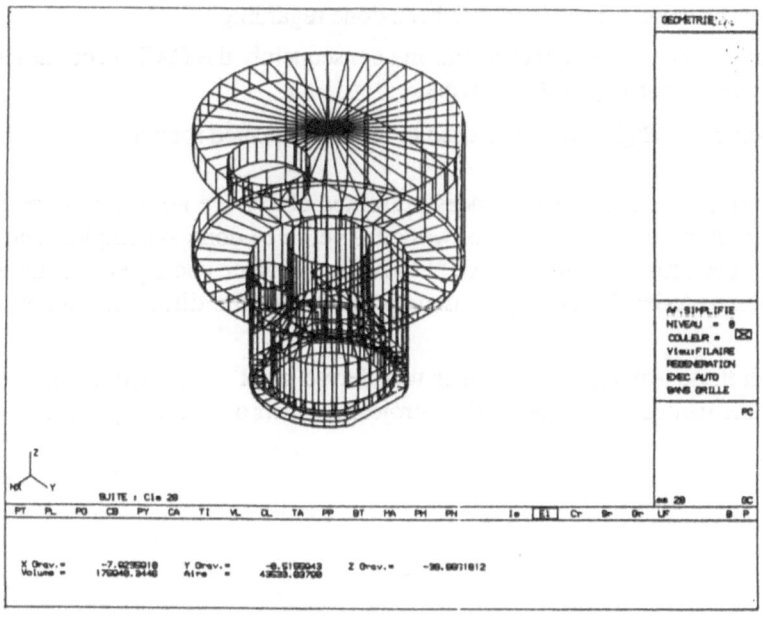

Fig. 8.4 STRIM model transferred in the CAD*I- ORACLE database

8.4 Experiences

One of the main goals for the CAD*I Database group was to develop a performant tool able to access CAD data supplied by every kind of CAD/CAM system according to the following aspects :

☞   The efficient storage and management by application programs of a large volume of data,

☞   The connection of these data with other types of data (e.g., administrative data).

The CAD*I database group considered that the utilization of a Database management system was the appropriate tool to respond to these aims, therefore one of the first tasks for the group was to choose a Database management system (ORACLE).

The second step for the CAD*I Database group was to develop a package of routines supporting the management of geometrical and administrative functionalities. This package has been specified and implemented in such a manner that every kind of CAD/CAM system can use this package to communicate with the CAD*I database. This successful result has been possible because the CAD*I Database group decided to specify and implement a standard package of routines written in standard FORTRAN-77 and supporting SQL portable statements to access CAD*I data stored. In addition, the specifications are available with the CAD*I neutral file format.

Concerning the CAD*I-ORACLE database interface implemented, CAD*I Database group experiences related to the performance of such a package becomes stronger in proportion as the project goes on.

Indeed, the latest results gained on the response times of the routines accessing CAD*I data in ORACLE are better, because a lot of work has been done regarding :

-    the optimization of the CAD*I data schema and consequently the CAD*I routines accessing CAD*I data stored according to this schema,

-    the optimization of the SQL statements used in the implemented routines.

Nevertheless, the first results and experiences presented above are not the best results CAD suppliers and CAD users can expect. A lot of work has yet to be done to complete and improve the capabilities and performances of the CAD*I database interface package of routines already implemented and used today in the applications programs of the different CAD*I database partners.

In accordance with these aspects, each partner will derive great advantage in future from the work and experiences gained during the CAD*I project for their own developments.

Addresses

BMW

Bayrische Motorenwerke AG
FS-30
Petuelring 130
D-8000 München 40 / FRG

CISIGRAPH

Compagnie Internationale de Services en Informatique
Les Bureaux de Parc la Griffon
590, Route de la Seds
F13127 Vitrolles / FRANCE

KfK

Kernforschungszentrum Karlsruhe GmbH
IRE/PFT
Postfach 3640
D-7500 Karlsruhe / FRG

UKA

Universität Karlsruhe
Institut für Rechneranwendung in Planung
und Konstruktion
Kaiserstraße 12
D-7500 Karlsruhe / FRG

References

A. Baer, C. Eastman, M. Henrion: Geometric modelling: A survey. Computer aided design 11(5) (Sept. 1979)

M. Benayoune, P. E. Preece: Methodology for the assign of databases for engineering applications. Computer aided design 18(5) (June 1986)

J. Date: An Introduction to Database Systems. Addison-Wesley (1981)

K. Dittrich, A. Kotz, J. Mülle, P. Lockemann: Datenbankunterstützung für den ingenieurwissenschaftlichen Entwurf. Informatik-Spektrum 8(3), 113-125 (June 1985)

W. Eberlein: CAD-Datenbanksysteme. Springer-Verlag (1984)

H. Grabowski, R. Glatz: Schnittstellen zum Austausch produktdefinierender Daten. VDI-Z, Bd. 128, Nr.10 (Mai 1986)

P. Lockemann, J.W. Schmidt: Datenbank-Handbuch. Springer-Verlag (1987)

M. Pratt : The CAM-I application interface specification. CAM-I Report R-86-GM-01, Vol I-II, Arlington (1986)

A. A. G. Requicha: Representations for rigid solids: theory, methods, and systems. Computing surveys 12(4), 437-464 (Dec. 1980)

A. A. G. Requicha, H. B. Voelcker: Solid modeling: A historical summary and contemporary assessment. IEEE computer graphics and applications 2(2), 9-24 (March 1982)

A. A. G. Requicha, H. B. Voelcker: Solid modelling: Current status and research directions. IEEE computer graphics and applications 3(10), 25-37 (Oct. 1983)

G. Schlageter: Transaktionsmechanismen für CAD/CAM-Datenbanken. (Datenbanktutoriantage Darmstadt), Fernuniversität Hagen (1987)

E. G. Schlechtendahl (ed.): Specification of a CAD*I neutral file for solids. Version 3.3, Res. Rep. ESPRIT, Proj. 322 Vol.1, Springer-Verlag (1988)

Shenoy, L.M. Patnaik: Data definition and manipulation languages for a CAD database. Computer aided design 15(3) (May 1983)

D. Steinbauer, H. Wedekind: Integritätsaspekte in Datenbanksystemen. Informatik Spektrum 8(2), 60-68 (April 1985)

S. Ulfsby, S.Meen, J.Oian: TORNADO: A DBMS for CAD/CAM systems. Computer aided design 13(4), 193-198 (1981)